Traveling the Power Line

Our Sustainable Future

Traveling the Power Line

From the Mojave Desert to the Bay of Fundy

JULIANNE COUCH

University of Nebraska Press | Lincoln & London

Publication of this volume was assisted by a grant from the Friends of the University of Nebraska Press.

Library of Congress Cataloging-in-Publication Data
Couch, Julianne.
Traveling the power line: from the Mojave Desert to the Bay of Fundy / Julianne Couch.
p. cm. — (Our sustainable future)
Includes bibliographical references.
ISBN 978-0-8032-4506-8 (pbk.: alk. paper) 1. Power resources—United States. 2. Electric power plants—United States. I. Title.
TJ163.25.U6C68 2013
333.793'20973—dc23 2012035997

Set in Sabon by Laura Wellington.
Designed by Roger Buchholz.

For Amy

Contents

Preface | ix

Acknowledgments | xiii

Introduction | xv

1. Of Megawatts and Meadowlarks: A Wyoming Wind Farm | 1
2. Angels and Monsters: A Wyoming Coal-Fired Power Plant | 24
3. Fission and Fishing: A Nebraska Nuclear Power Plant | 49
4. Solids, Liquids, and Gases: A Texas Gas Field | 74
5. Homegrown Revolution: An Iowa Biomass Research Facility | 99
6. Journey a Little Way into the Earth: A Utah Geothermal Plant | 123
7. Water, Water, Everywhere: A Kentucky Hydropower Plant | 143
8. Don't Let the Sun Go Down . . . without Capturing Its Energy: A Nevada Solar-Thermal Power Plant | 166
9. Harnessing the Moon: A Maine Tidal Power Project | 188

Afterword | 212

A Note on Sources | 213

Preface

This book tells the story of my travels to various sites of electrical power production across the United States. The trips were made between spring 2008 and spring 2010, with Laramie, Wyoming, serving as my home port. Expedition bulletins that start each chapter acclimate fellow travelers to conditions there.

In Wyoming there exists almost every sort of power plant or form of electrical energy production described in these pages. However, because the population of Wyoming is only around half a million, word of its role in the energy game has not spread nationwide. For that reason, I wanted to visit places where natural resource development and energy production were not typical dinner-table conversation, as they are in Wyoming. In that way I could learn firsthand not only about hydroelectric power, for example, but about how hydroelectric power shapes the lives and landscapes around it.

I chose to visit a wind farm and a coal-fired power plant in Wyoming because these forms of power production are virtually synonymous with that state. Starting from my home base gave me ways to think about approaching my visits to other types of power less familiar to me. The nuclear plant in Nebraska is one of the westernmost plants before the arid Great Plains, where that water-intensive power source is rare.

I could have found a power plant using natural gas as a fuel source, but it would have been much the same experience as my visit to the power plant burning coal as fuel. So I chose to visit Texas and its vast underground Barnett Shale to see firsthand how shale gas was extracted via drilling and fracturing.

Iowa seemed a logical place to see the sorts of possibilities that biomass materials such as corn stover and algae hold as fuels. Also the southern Utah desert is one of the prime areas for geothermal resources nearest to the surface that makes exploration and exploitation of that resource possible.

In the course of my travels I took a public tour of the Hoover Dam on the Nevada-Arizona border, not far from Las Vegas. However, I chose to focus my exploration and research on the Tennessee River Valley, a historically important spot in the development of electrical resources in rural areas of the south. That's how I found Kentucky Lake Dam.

Solar energy in California leads the development pack, but the site I selected in Nevada outside of Las Vegas allowed me to visit both the collection site and to see the connected power plant, fueled by sun-heated fluids.

Visiting the tidal power development site in Maine was a simple choice. No place else developing this sort of technology was anywhere close to having something to show me.

One element that might seem a logical topic for inclusion in this book is oil. Discussion of natural gas' first cousin is included in chapter 6. My decision not to focus specifically on oil was motivated by the fact that oil is used primarily as a transportation fuel. True, many households, especially in the Northeast, use home heating oil to stay warm in winter. But the focus of this book is on fuels used in power plants to create electricity, the sort of thing we use daily in our houses and in industry and agriculture. A short list of excellent resources on oil and other related matters can be found in the back of this book.

One of my goals was to listen to the voices we all have access to every day speaking about electrical energy, whether they are motivated by issues of climate change, national security, or conservation. Rather than pitting these voices against one another, as happens too often in our national discourse, I prefer to blend the smart and sincere positions of scientists, engineers, policy advocates, environmental activists, industry experts, and the folks who work in or live around various sites of energy production. By hearing them, as well as seeing in person what they are talking about, I hope to convey a fuller sense of something as fundamental to our lives as electrical power. In this way I hope to show how our current technologies exist together and what might change as we travel down the power line.

Acknowledgments

This material is based in part upon work supported by the University of Wyoming School of Energy Resources through its Matching Grant Fund Program. It was also supported in part by funding from the Wyoming Arts Council and the University of Wyoming Department of English.

Thanks to those who made valuable suggestions to help make this book worth the read, including Paul Woolf, Peter Ozias, Bill Markley, John Freeman, Laurie Milford, Jeffe Kennedy, Doug Couch, RoseMarie London, Bill Bishop, Julie Ardery, Randy Rodgers, and Ronald Hansen. Additional thanks to Brian Hayes and John McPhee.

The Department of Energy's Energy Information Administration's guidance and prompt replies to my numerous inquiries were always helpful, professional, and prompt. Without them there would literally be no book.

Finally, although I made every effort to write about various forms of energy production clearly and accurately, and sought outside experts to read chapters, technical errors may exist. If errors of fact exist, they are unintentional but are mine alone.

Introduction

"Good luck in the English Dept. Hope this isn't too confusing."

With these words, a staff member at the U.S. Department of Energy's Energy Information Administration (EIA) correctly pegged me as a have-not in the world of technical knowledge. I'd e-mailed her for help interpreting some data on the EIA website. The data was about electricity consumption versus production, and I was admittedly confused about the relationship between the two as presented in a particular table. I signed the e-mail message with my name, taglined "University of Wyoming, Department of English."

I felt pretty smart for detecting the nuance between energy production and consumption and I admit I wanted to give my perceived smartness some heft by tacking on a few credentials. After all, English departments are better known for producing and consuming poetry than power. The first part of the e-mailed reply gave me a link to another table where my question was answered directly. But the second part, wishing me luck, I interpreted as being delivered with bemusement. Or maybe the line was meant sincerely and in the vacuum of e-mail communication was devoid of context or tone.

That I immediately leapt to the conclusion I did perhaps

speaks to my own embarrassment over how little I know about some very important things. There I was, a generalist trying to pass as someone who belonged in the world of gigawatts and generators. But I'd undercut my own ethos. I might as well have signed my name Julianne Couch, PUBW (Professional Underwater Basket Weaver.)

A 2011 Harris Poll revealed that I am not alone in my ignorance. About 12 percent of Americans indicated they were "very" knowledgeable about where electricity comes from. About half thought they were "somewhat" knowledgeable. Until I started this power research I'm not sure I could have even passed myself off as a somewhat-er. Even now, after personally seeing power produced from the energy of the wind, the sun, and the inside of the earth; after witnessing the potential of natural gas, pulverized coal, and the stover of corn; after seeing atoms split apart, rivers harnessed and tides captured, it still takes all my powers of cogent concentration to explain the industrial process to others. I could never produce electrical power myself or be of much use on a team of folks who do.

Where I live in Wyoming, energy extraction and export is the marrow in our bones, affecting everything from whether we can afford new schools and highways to how our social infrastructure is influenced by itinerant energy workers. It is impossible to open a newspaper, listen to local radio and television, or drive down our byways and back ways without noticing energy development and environmentalism tugging each other from opposite sides of the same rope. Although some claim these two positions are not mutually exclusive, the fact is that arguments over energy issues sound their alarms so loudly they become background chatter, without meaning. If a person like me, living in one of the richest states for reserves of both fossil and renewable energy, doesn't quite have a grasp of the issues involved, odds are people in other locations are at an even greater disadvantage.

I have learned the basics on this power tour: that energy

comes from fuel sources, and power is a measure of the fuel source used for a given amount of time. Some fuels provide plants with baseload power—they run all the time—while others are intermittent. The basic unit of electrical power is the watt. One kilowatt-hour can power one average-sized home for one hour. A megawatt is one million watts of electrical power. That is enough to power about a thousand houses for an hour. A single wind turbine may have a peak capacity of one megawatt, and if it operated continuously at that capacity for every hour of one year, the energy it would produce would be 8,760 megawatt-hours. But it doesn't produce that much because the wind doesn't blow all the time; therefore, its capacity factor is closer to 20 percent. Capacity factor refers to the ratio of power actually generated to the maximum potential generation, expressed as a percentage. In contrast to wind, a coal-fired power plant of 500 megawatts operates at about 80 percent capacity. That's why it takes 2,000 wind turbines to replace one 500-megawatt coal unit. That's a mighty big wind farm.

Being able to recite these facts does not mean I am an expert. Thus, I can't predict whether another rash of nuclear power plants will be built in this country after the disastrous nuclear episode in Japan in 2011, or to what extent carbon emissions will be captured from fossil-fuel plants and whooshed into underground cavities to force out more gas. I don't know whether the desert tortoise in the Southwest will thrive surrounded by concentrated solar power plants. I couldn't say with authority whether dams in the Pacific Northwest should be decommissioned to allow wild fish to migrate. And I still don't know much about the complex and obtuse regulatory environment in which energy policies are shaped. But I'm certain that I now realize how much these questions matter and have a personal connection to them, and I can approach them more thoughtfully than I once did.

Perched as we seem to be on the cliff of a great energy

evolution, it is important to be forearmed with awareness. Yet it is difficult to know what sources of information to believe, whose agendas to trust. Power companies research and develop energy sources, including renewables, while saying very little about conservation. The "people of the oil and gas industry" run advocacy ads during the evening news, explaining that we have all the resources we need in this country to "free ourselves from dependence on foreign oil." Environmental activists argue there is nowhere we can drill for oil or pipe natural gas or place wind farms that is worth putting wild creatures or pristine landscapes in peril. Little is said about the pressure of an ever-growing number of people in the world and in this country, which strains all our resources, not just energy alone.

Like most of us, when I am armed with information, I am better at analyzing arguments put forth by very smart, persuasive people who appear to believe in what they are saying. Show me a nuclear plant manager who says we also need wind farms in this country, and I'll show you someone who is listening, not just talking. Show me a coal-plant worker concerned that the reservoir dammed to supply the plant with water is clean for his weekend fishing, and I'll show you someone credible to speak up if things get sloppy. Show me an environmental group that acknowledges the impracticality of shuttering all coal plants tomorrow, and I'll be more willing to do what it takes to weatherize my house and conserve energy.

The bottom line is not whether but when our country will be moving to more renewable fuels. In fact, projections by the EIA forecast a slow but steady move toward renewables and fuel options lower in carbon emissions. The hows of this shift will be driven by policymakers, scientists, engineers, and entrepreneurs as states develop their renewable energy goals. The whens will be driven by us, the consumers.

The U.S. Census Bureau tells us that we started 2010 with 308,528,525 of us, an increase of 9.7 percent since 2000. To satisfy this ever-growing population we'll want more of everything people generally want: houses and cars, roads and sewers, movie theaters and popcorn, wedding dresses and diapers. We will need commercial-scale agriculture and industry to feed, clothe, and supply us; therefore, we'll need fuel to fire our power plants to get those jobs done.

Let it not be lost that we'll also value low-density and natural spaces to sustain our sense of peace. We won't want to poison our air and land and water with industrial waste, and we might not want to look at spinning wind turbine blades on every hilltop or coastal landscape.

All of these "wants" are not the same as needs. Do we need to cool our homes to 70 degrees when the temperature outside is 90? Do we need that second refrigerator in the basement to chill our surplus soda pop? Do we need the home theater systems that consume so much electricity? These are issues to visit on the individual scale, but they are magnified exponentially when considering the industrial and agricultural practices of our nation.

I'm not an expert. But there is a cumulative effect of general firsthand experience, of walking a ridgeline among wind turbines, of standing just a few feet from the used fuel pool of a nuclear plant, of strolling along the top of the geothermal field warmer than the Utah desert that surrounds it. Brian Hayes, author of *Infrastructure: A Field Guide to the Industrial Landscape*, explains that it is important to have at least a rudimentary understanding of the way things work. "Without a sense of how materials and energy flow through an industrial economy, you miss something basic about the world you live in." And he argues that there is a practical value in knowing how things work. "People who have never seen a power plant, who know nothing of how it works, who have never met anyone who works there, are poorly equipped to judge the relative merits of nuclear and coal-fired technologies, or to seek alternatives that might allow us to dispense with both. To make good decisions about such issues, citizens need to get better acquainted with the technological underpinnings of their own communities."

We are in a new energy crisis while we choose between the fuels that got us where we are and the ones that will get us where we are going and help us lessen our dependency on fuels of all types. What used to be a fairly straightforward formula of what fuels had what uses has recently become more complicated. For example, automobiles were once powered by petroleum products. Now they can also be powered by electricity, biofuels, or other sources. These other sources aren't environmentally pristine just because they aren't petroleum, although their advocates don't often mention the down sides. In fact, almost all electrical power has traditionally come from the burning of fossil fuels. These days it can also come from generators burning 100 percent used vegetable oil. It can be hard to keep up.

According to that 2011 Harris Poll, only one in five U.S. adults are "very" interested in keeping current about energy issues, although 53 percent say they are "fairly" interested. I'll put my faith in those fairly interested folks. Only a few of us can be experts. But all of us can look beyond the power switch and see what goes on behind the wall. If we discover enough to ask better questions, perhaps we'll understand the price we pay each time we flick a switch. Somewhere, someone throws another chunk of fuel on our insatiable national fire.

Say, maybe it isn't that confusing after all.

Traveling the Power Line

1

Of Megawatts and Meadowlarks
A Wyoming Wind Farm

JANUARY 1

The Brees reporting station at the Laramie Regional Airport registers 32 degrees Fahrenheit, with steady twenty-five-mile-per-hour winds, gusting to thirty-five. The airport sits at an elevation of 7,266 feet. About seven miles east, my home sits on one of the highest hills in Laramie, Wyoming, at 7,333 feet. The high and low temperatures these last few days have been seasonal, which, after the very cold temperatures in December, seem almost balmy.

This morning I got out of bed ahead of my New Year's Eve–addled neighbors and walked my small dog. Though Archie's pointy ears blew backward as we faced into the wind, his four feet stayed on the ground, two at a time. I timed our outing to avoid the even higher winds that would develop in the afternoon. Powering up my laptop after returning home from our walk, I learned from the National Weather Service website that a high wind advisory had been issued for later today.

> A high wind watch remains in effect from 5:00 p.m. MST this afternoon through Friday afternoon. There is the potential for very strong west winds to return again tonight and much of Friday . . . and a high wind watch remains in effect for those times. A high

wind watch means there is the potential for a hazardous high wind event. Sustained winds of at least 40 MPH . . . or gusts of 58 MPH or stronger may occur. Continue to monitor the latest forecast.

Wyoming is one of the windiest spots in the nation, with southern Wyoming, where I live, heading the stats. Paradoxically, blowing winter winds warm the air temperature and melt the packed snow and ice that cover my street. When the wind stops we are in trouble: that's when the cold settles into every nook and cranny, making us as chilly as a chest freezer.

Long ago, before I moved to Wyoming from Missouri, a woman I met who once lived in neighboring Colorado assured me that "all people in Wyoming are crazy"—driven mad by the perpetual wind. It wasn't just her opinion: there's a 1928 silent movie called *The Wind*, in which Lillian Gish slowly goes mad in a dusty town along the Sweetwater River, in south central Wyoming. And a favorite local sight gag involves a heavy gauge chain being blown perpendicular to a post. Attached to the post is a sign identifying the strength of wind as indicated by the angle of the chain. According to the sign, a reading of zero degrees indicates the wind sock is broken and the meteorologist should be notified. A thirty-degree angle indicates a fresh breeze. Forty-five is a gentle zephyr. Sixty alerts us a hurricane is in the area; seventy-five and those in the area should beware of low-flying trains. At ninety degrees, the sign proclaims: Welcome to Big Wonderful Wyoming.

In the winter, wind sends snow blowing in horizontal curtains or snaking across highway pavement where it melts just enough in the high country sun to leave behind a sheet of ice. As the sun sets, that ice lurks beneath the snaking snow forming black ice, treacherous to motorists who do not detect its presence until they hit a patch and go spinning out of control across lanes of oncoming traffic. Wind causes the same conditions in spring, when in much of Wyoming and other mountainous western states the snow falls heavy and wet.

Most areas of the state traversed by I-80, Wyoming's main east-west drag, are breezy regardless of season. Interstate 25, which runs north-south in the eastern section of the state, is not exactly calm, either. The Wyoming Department of Transportation (WYDOT) searches for ways to reduce what are known as blow-overs on the interstates—incidents in which vehicles simply topple when high winds hit them just right, or just wrong. WYDOT reported that between 1998 and 2007, high winds caused 1,544 crashes of trucks and other vehicles on Wyoming highways, or an average of about 150 a year. These are just the crashes due to strong winds; the number does not include accidents to which icy roads, blowing snow, or other dicey weather conditions contribute. Of those wind-only crashes, 623 led to injury, and 19 were fatal.

Although the whole state is breezy, communities with average sustained annual wind speeds below ten miles per hour take smug pride in having less of it. Those towns like to consider themselves the state's garden spots because they clock in with average sustained annual wind speeds of only six or seven miles per hour. Sustained winds in towns most notorious for high wind, such as Casper, Cheyenne, Laramie, and Rawlins, clock in at around twelve miles per hour. Although twelve miles per hour on average doesn't sound like much, factor in the number of hours the wind relaxes, such as after the sun goes down, and you've got enough breeze to fly a cadre of kites. And that's wind speeds in town, not high up on ridges or mountain tops. Go to a place like that, then stick a long moist finger two hundred feet up in the air around the height of a wind turbine, and you will notice that speed has doubled.

Wind has long been used by clever humans to sail ships, grind grain, pump fields, and dry laundry. Henry David Thoreau wrote about the power of wind in "Paradise (To Be) Regained": "Here is an almost incalculable power at our disposal, yet how trifling the use we make of it! It only serves to turn a few mills, blow a few vessels across the ocean, and a few trivial ends besides. What a poor compliment do we pay to our indefatigable and energetic servant!"

Wind has also been used to create electricity, although until recently widespread use for that electricity was not practical. But now with a burgeoning infrastructure of transmission towers and power lines, wind is touted as a practical renewable source of energy, more abundant and cleaner than coal. In Wyoming, and in other windy locations in the United States and around the world, commercial wind farms are sprouting like corn. The tall towers of a wind farm reach their long blades hundreds of feet into the sky to greet the breezes that power the generators that feed electricity into a transformer and eventually into the power grid and on to consumers across the country.

Wyoming has undergone several boom-and-bust periods in its history, mostly as a result of the extraction industries. When material such as oil, gas, uranium, and coal are removed from the ground and exported to provide power in other states, that's known as extraction. When the various companies that do the extracting take the materials away, they pay a severance tax because they've severed those materials, never to return them.

Wyomingites live large during those boom times, when severance taxes support our schools, roads, bridges, and other items paid for in many other places by state income taxes, which we do not pay. But when for various economic reasons companies slack off on the extracting and the severing, people in Wyoming start to feel the pinch. Schools and motels and convenience stores and recreation centers built to handle the boom find themselves with hardly any students, guests, shoppers, or recreationalists. In local parlance, that's known as a bust.

Places in the country long on people but short on energy resources have become used to the riches of Wyoming and other energy-exporting states. So when the rest of the country sees fuel costs rising, it makes a certain sense to say, "Why don't we just drill, baby, drill, for more in Wyoming?" In most cases, the answer both inside and outside of Wyoming has been, "How deep?" Just as dizzying as the boom-and-bust economy has been for Wyoming's state budget, Wyoming's environment has also taken a hit. Wyoming once was renowned for clean air, open spaces, plentiful wildlife, healthy streams, and breathtaking scenery. More and more, with booms mostly dependent on the natural gas industry, it appears to many of us that the air and water have been sullied and wildlife have become skittish or downright threatened by human activity. Sure, we received money to fund long-deferred projects. And students have a better chance to attend college due to an in-state scholarship funded by those dollars. But we've paid a pretty tough price.

Enter wind.

At last, a chance to develop an energy product without severing it. Much to the chagrin of some people on those gale-force winter days, Wyoming has plenty of breeze to go around. There are those who wouldn't mind if all these new wind farms would deflect some of the breezes south into Colorado. But I began to wonder if renewable energy, specifically wind power, could meet all our energy and economic needs and yet not damage the environment. I'd noticed a wind farm under construction along the interstate about an hour from my home, but it was difficult to determine how to arrange a visit. There was no sign of an office, or even a shop, where one might inquire about getting a tour. So I picked up the phone and made a call to my local power company. One week, several phone calls, and many e-mails later, I made my way up the chain of command to Jeff Hymas in the external communications office at Rocky Mountain Power, based in Salt Lake City, Utah. After a bit of confusion, I learned that's how the company is known in Utah, but in Wyoming and other states, it operates as PacifiCorp. Hymas told me that public tours of the wind farm were not allowed, but that I could visit there as a journalist, accompanied by wind-farm staff, at a time of mutual convenience.

He set up a date and time for me, even sending me detailed directions to the small structure where wind-farm employees worked. I'd never noticed the building before because it was on a two-lane side road, rather than the interstate. Both before my trip and after, Hymas made sure I had all the information available about the company's wind development plans for the area, as well as the "clean" nature of the energy produced at Foote Creek Rim.

The Foote Creek Rim wind farm is Wyoming's first commercial facility to generate electricity from wind. It was originally developed by SeaWest, which was then purchased by global energy conglomerate AES Energy. PacifiCorp is one of several wind-farm developers that AES hosts at the location. PacifiCorps's site began commercial operation on Earth Day, April 22, 1999, starting with sixty-nine towers.

PacifiCorp has constructed several additional wind energy projects in this area of Wyoming, including the Glenrock, Rolling Hills, and Glenrock III projects in Converse County, located in eastern Wyoming, and the Seven Mile Hill and Seven Mile Hill II projects located near Foote Creek Rim in Carbon County. These projects were completed and brought into service in December 2008 and January 2009. The company's High Plains wind project was constructed on land spanning the border of Carbon and Albany Counties, and came online in 2009. These projects will add to the 2,000 megawatts of renewable energy resources the utility plans to add to its generation mix by December 2013.

But it isn't enough simply to create wind power. There also have to be transmission lines to carry the electricity to customers. For that reason, PacifiCorp, together with Idaho Power, is engaged in a project dubbed Energy Gateway West. Once completed, transmission towers will string together power lines stretching from eastern Wyoming to western Idaho.

I considered taking the route of the future transmission towers to tour Foote Creek Rim, located about forty miles to the west of Laramie. That would have taken me along I-80. The weather around the interstate isn't bad in summer, though in winter it lives up to its local nickname, the Snow-Chi-Minh Trail. All year-round the interstate hums with tractor trailers and other large conveyances hauling bulldozers, pieces of pipe as big as my car, and other colossal gadgets to places in Wyoming where energy companies work virtually around the clock. Adding to that mess are a jumble of summer tourist motor homes and motorcycle packs, prompting me to choose the longer but scenic back way north to the town of Rock River, then west through the ranching communities of McFadden and Arlington.

As I approached from a distance of a few miles, it appeared that a white-clad Edward Scissorhands was lying on his back across Foote Creek Rim, lazily waving his blades in the breeze. In fact, that body of swiveling blades that reaches about five miles across this rim is the wind farm, with turbines up to two hundred feet tall slicing the sky. Jeff Hymas had arranged for me to tour the wind farm with Tony Kupilik, team leader for project operations there. Kupilik welcomed me to a modest control room next to his office and showed me how the turbines are run by computers. Operators can turn turbines off when high wind conditions make it necessary, either from the farm or from corporate offices in California, or even from Kupilik's home in Laramie. He showed me how to read data on the monitors, one segment for each turbine and for the two meteorological towers. Wind speed is measured in meter-seconds, which is roughly half of the miles-per-hour measurement. With average wind speeds of twenty-five miles per hour, the monitors show that Foote Creek Rim is one of the windiest spots in the United States.

Next, Kupilik pointed out a topographical map indicating the steepness of the Foote Creek Rim and the width of the flat plateau, with turbine positions superimposed. The land is owned in a checkerboard combination of private owners, the state of Wyoming, and the Bureau of Land Management (BLM). The public lands are landlocked by private lands, so no one may legally access the site without permission to be there. That's why no one is wandering around the wind farm unaccompanied.

Kupilik showed me a schematic drawing and gave me a quick overview of how wind turbines work. He explained that wind turbines create power when wind blows across their blades, which are attached to rotors. Air forces the blades to spin, which drives the shaft to which the rotor is connected, which creates the electricity. In some ways it is similar to the action of a hydroelectric generator found on river dams. But because air is less dense than water, the blades of a wind turbine must be much larger than those in a hydroelectric generator.

I'm no mechanic, but it was a comfort to hear him use some terms I was familiar with: blade, rotor, tower, brake, gearbox, generator. I felt armed with the basic comprehension I needed to head up to the rim and see 134.7 megawatts of electricity generated per hour. Kupilik reminded me that one megawatt is enough to power about one thousand homes. He showed me to the Dodge company pickup and commiserated with me about the difficulty of climbing into a truck that sits high off the ground and lacks running boards. "We get so much snow up here in winter running boards would just get knocked off," he told me. As I gathered the camera and notebook that I'd heaved onto the seat ahead of me, he glided the truck up a gravel road and through a gate, taking us down an access road through private property. We could hear a western meadowlark through the open window, trilling its greeting.

As the truck climbed toward the rim, we took in views of the dark green Medicine Bow National Forest just south of the interstate. Always-snowcapped Elk Mountain loomed in the southwest sky. The wide-open Laramie Basin to the east and north opened up below us. Kupilik pointed out some lakes where he used to fish as a Colorado kid who summered with his grandfather in Wyoming. He spent much of his time in Rock Creek Canyon, an area we could make out from the truck. Kupilik is about fifty years old, and when he was growing up, this ridge was a short-grass and wildflower plateau during the summer, not bristling with spinning rotor blades on tall towers like it is today. He isn't disturbed about the change in scenery. "I like the way they look," he said about the white turbines. "Their being here doesn't bother me at all—I think they look cool."

Summer is a "low wind season," Kupilik explained, so that is the time they take turbines off-line to perform routine maintenance. Kupilik pointed out a turbine whose white UV reflective paint was marred by some oil seeping from the rotor. "We're going to get that cleaned up. We're pretty respectful of the environment up here," he said. Although summer is a busy time for maintenance, it beats working on the turbines during winter. Not only does the icy wind blow nonstop, and the snow fall almost as vigorously, the towers themselves are extremely chilly. Like lighthouses, the towers have a small door at their base and an interior ladder that workers climb to reach the business end at the top. "If the turbine has been running, it can stay fairly warm up there, but after a few hours it gets very cold inside," explained Kupilik in a voice that said he'd spent a few too many repair sessions in just those conditions. "It's pretty miserable up here in winter."

Kupilik has worked for AES for several years. Once employed at a coal mine at nearby Hanna, Wyoming, he needed a job after the mine shut down. He had no training in wind turbine maintenance at the time but had enough mechanical know-how that he was hired. He learned on the job.

Now Kupilik's experience has landed him on the advisory board for the Laramie County Community College (LCCC) program established to train and certify wind-turbine technicians. LCCC, in Cheyenne, Wyoming, is one of a handful of programs in the nation to offer a wind-energy training program, in which students can earn an associate of science degree with a concentration in wind energy. Mike Schmidt, an industrial electrician by training, is LCCC's Wind Energy Technical Program director. He came to Wyoming from Iowa Lakes Community College in 2008 after establishing a successful wind-energy training program there. He said LCCC's first class had twenty students, about half of whom were from out of state. The students were a mix of ages, from traditional college age up to age sixty. Several of them already had bachelor's degrees in other fields. Schmidt said they entered the program because, like him, they believe in wind energy. They wanted a career in a field that not only offers a good wage but can make a difference. Schmidt believes that global warming is proven by science and that human activity has added to it. "If we have the technology to reduce global warming, we should do it," he said.

Both the technology and the regulatory environment in Wyoming are combining to make wind energy a major player in nontraditional energy development in the state. Wind energy capacity in Wyoming increased from 288 megawatts in 2007 to over 1,400 megawatts at the end of 2010. In addition, nearly 7,900 megawatts of new wind projects are in some stage of development. Capacity means how much power wind farms are delivering onto the grid. Capacity also means what it sounds like. If the wind blew everywhere, all the time, at an optimal speed for wind power production, power could be produced at the full capacity of the existing wind farms. But wind, being capricious, comes and goes as it pleases. That is why power companies don't just plop wind farms anywhere. The ins and outs of wind-farm siting were explained to me by Jeff Hymas and his boss, Dave Eskelsen. First, developers look for places where the wind is dependable and constant. Then they look to see whether those places are already in close proximity to existing transmission lines. If they have to sink money into infrastructure to move the electricity, the costs rise for them and for their customers. Also important is the ground itself.

It needs to be stable and fairly level. Foote Creek Rim, for example, is constantly windy and reasonably level. And the power infrastructure already in place along I-80 makes clustering other wind power projects there attractive.

On the other hand, environmental groups, government agencies, and various collectives of stakeholders have additional, sometimes contrary ideas about ideal sites. For example, the Laramie-based Biodiversity Conservation Alliance (BCA) applauds the use of wind power but argues for siting wind farms where they will be least disruptive to humans and wildlife. They note the unprecedented surge in wind-energy development proposals both for public lands, such as forests, grasslands, and basins, and for privately owned land.

The BCA undertook an analysis of wind-farm siting, specifically in Wyoming. In 2008, they released the results of their study in a document titled *Wind Power in Wyoming: Doing It Smart from the Start*. BCA Executive Director Erik Molvar, a wildlife biologist and the author of the report, argues that Wyoming should develop wind energy "in a way that protects open spaces and native ecosystems and is an asset to local communities rather than a disruption." The report was endorsed by the World Wildlife Federation, the Wyoming Wilderness Association, the Earth Friends Wildlife Foundation, the Western Environmental Law Center, the Wild Utah Project, the Sierra Club, and Californians for Western Wilderness. The report includes a color-coded map depicting where wind development is appropriate, and where it is not, based on a variety of conditions.

In addition to the conditions for wind-farm locations most valued by power companies, these groups advocate caution in siting large commercial wind farms near landscapes such as national parks, wilderness areas, and national forests. They argue for the protection of raptors and bats, conservation of sage grouse and big game animals, and general stewardship for sensitive wildlife. Finally, they point out that aesthetics should be valued. What most of us think of as scenery has become known in this conversation as the "viewshed." Many people do not want an aesthetically pleasing viewshed cluttered with twirling turbines.

Eskelsen notes that the environmental concerns voiced by the BCA often arise when a power company proposes any sort of major project. Power companies are accustomed to entering into lengthy discussions with stakeholders such as citizen groups and governmental entities, in which concerns are expressed and addressed. Seldom does one group get everything it wants, he said, but the conversation is part of the process. "Utilities aren't welcomed with open arms," he acknowledged. And even though few people would be satisfied with a modern life without commercially available electricity, Eskelsen said, "It is very difficult for a utility to build infrastructure of any kind."

In addition to the BCA's recommendations, another group in Wyoming has developed a set of guidelines for siting wind farms. The state legislature created the Wyoming Infrastructure Association (WIA) in 2004, given the task of "diversifying and growing the state's economy through the development of electric transmission infrastructure." They see to the planning, funding, building, maintenance, and operation of interstate electric transmission and related facilities.

Aaron Clark is an adviser to the WIA. He lives and works in Wheatland, in eastern Wyoming, one of those areas where the wind capacity is robust, to say the least. He has worked as a consultant in the energy industry for more than twenty-five years and was the individual responsible for creating the state's map for appropriate wind energy development. Both the WIA and BCA maps use the color green to show where wind development is appropriate and pink to show where it is not. Both maps use similar criteria for determining where wind development is not appropriate.

According to Clark, both maps demonstrate that critical winter range for big game animals, sage grouse nesting areas, and highly sensitive visual areas, along with a myriad of other elements, should be off-limits. The BCA would also like a five-mile buffer zone between wind development and towns. Clark says he would prefer to leave that issue to local planners rather than to the state. Bottom line, the maps are in general accord and the "message is the same," Clark said. Molvar agrees. It isn't every day that conservation and industrial or governmental organizations agree with one another, without first undergoing court-mediated friendships.

So far, most wind-development projects in Wyoming have been on private land, where the permitting regulations are limited to those established by the state. The developers of any project on federal land, including those managed by the BLM, must also follow guidelines of the National Environmental Policy Act of 1969, which include environmental impact studies. Developers on federal lands must also adhere to the National Historic Preservation Act of 1966, specifically when developers identify cultural artifacts on the land that might make the area eligible for inclusion on the National Register of Historic Places. Significant Native American cultural artifacts were identified at Foote Creek Rim during the course of these studies.

There are approximately six hundred manmade rock features in the area, mostly on the edge of the rim. Walt George, national project manager for the BLM's office in Cheyenne, explained the sequence of activities this discovery prompted. First, the BLM conducted a literature search to determine what might already be known about historic Native American activity in the area. Then they consulted with leaders from the Northern Arapaho and the Eastern Shoshone tribes who share the Wind River Indian Reservation in Wyoming, to determine what those groups might know about their ancestors' use of the area. George said that leaders from the tribes examined the site and interpreted the meaning of some of the rock cairns and other rocks arranged in various patterns. Tribal members reported that some of them were used as navigation tools for nomadic groups to reach important locations around Elk Mountain, considered a sacred site.

Bonni D. Bruce, supervisory archaeologist with the BLM's High Desert District field office in Rawlins, Wyoming, said in addition to the Eastern Shoshone and the Northern Arapaho, other groups had ties to the area as well, such as various Sioux bands and the Southern Cheyenne. "Most tribes don't really know about these types of sites on public lands until we consult with them. They do recognize, however, features that were commonly created by their ancestors when we take them to sites. We've had tribes look at a cairn, for example, and say that it wasn't their ancestors who created it."

After the cultural features were discovered and their significance assessed, the developer, SeaWest, redesigned certain roads and wind-farm infrastructure to create minimal disturbance to the area. In addition to addressing cultural concerns, the impact on area wildlife had to be studied before construction could begin.

"At first people were concerned the wind farm would have a negative impact on wildlife up here, especially the mountain plover," Kupilik told me. "It seems to me, though, that they are doing fine." He's noticed that predators such as coyotes avoid the area more than they once did, perhaps because of the human presence. But he thinks the birds, small mammals, and antelope, are flourishing. "You should see this place during hunting season—it's crazy with antelope," he said. It makes sense that antelope would head to places off-limits to hunters—no one is shooting at them on Foote Creek Rim.

Impact on wildlife is of primary concern when determining wind-farm sites. In some of the earliest wind projects in the United States, such as the one at Altamont Pass in California, wind towers were concentrated in areas frequented by raptors, thousands of which were killed annually by collisions with whirling rotor blades. According to Walt George, the mortality numbers from that site were all that were available to use as a comparison for what would happen at Foote Creek Rim, simply because there were few other large-scale wind farms. But the environments of the two locations are quite different, he said. The Altamont Pass site is surrounded by areas of dense development in the San Francisco Bay Area. Therefore, the underdeveloped area at Altamont Pass was the chief place where birds concentrated. But at Foote Creek Rim, as in much of Wyoming, there is little urban development to force birds into unnatural concentrations. Because the BLM knew the numbers at Altamont were not a good comparator, they initiated a raptor survey at the Wyoming site.

The National Academy of Sciences determined in 2006 that an average of three out of one hundred thousand human-caused bird or bat deaths were due to wind energy activity in the United States. However, migration pathways for birds and bats can be highly traveled at certain times and so mortality numbers at specific sites can rise much higher. For two years before developers constructed the Foote Creek Rim wind farm, wildlife experts studied the way birds used the area around the rim. They learned that bird activity concentrated around the lee side of the rim, where birds sat on rock formations or soared in the thermals of the gently rising land. As a result of their findings, SeaWest located the turbines away from the edge of the rim. Once the turbines were in place, consultants studied avian mortality levels at the site for five years after that. Their research was reviewed by a technical committee including Wyoming Game and Fish and the U.S. Fish and Wildlife Service. At the end of that period the committee determined that mortality levels were in keeping with avian mortality in general for the area, and that there was no reason for further study.

Governmental and environmental groups have both worked to prohibit the placement of wind turbines in locations where wildlife could be disturbed or harmed. Walt George of the BLM says most stakeholders agree that the ideal spot for siting a wind farm is on private, rather than public, land. That's because on ranchland, for example, some habitat disturbance may have already occurred, and wildlife using the area may already be habituated to human activity. Erik Molvar, with the BCA, categorizes some views on public land as "sensitive landscapes." According to his report, "Wyoming is known throughout the world for its iconic western landscapes. Many of these, like national parks, wilderness areas, and wilderness study areas, have been placed off-limits to industrial activities by federal law or regulation. Others, such as roadless areas and BLM Areas of Critical Environmental Concern, have limited protective designations, which would tend to frustrate the timely development of wind projects and might preclude them in some cases."

Foote Creek Rim is far away from national parks. In fact, casual observers would not notice anything unique or special about the windswept hills of Wyoming's basin-and-range country, where it is located. But as Tony Kupilik drove me around the project on that summer day, along private roads far off the interstate, I renewed my appreciation for the location's sage and rabbit brush prairie. I could see why humans had been attracted to the area for so many centuries, surrounded as it was by bountiful mountains and plenty of water. Aside from the few ranch fences and the wind farm itself, the view of Elk Mountain from Foote Creek Rim has remained much the same as it was for the Native Americans historically traveling through the area.

After Kupilik and I took in some of these grand views from the truck, we passed the substation that receives power from all the turbines. Regardless of their height, he explained, the turbines each produce 600 volts. A step-up transformer converts that low-voltage electricity to a high voltage so that it can travel more efficiently for long distances down power lines, he told me. To start that process, a transformer next to each turbine steps the voltage up to 34,000 volts. Power enters the substation through a buried cable and is boosted to 230,000 volts, which travels down transmission lines to another substation in Hanna, about thirty miles away. Electricity traveling along power lines encounters friction to a degree that depends on the type of transmission lines it is on. The EIA estimates that loss of efficiency at less than 7 percent. They measure efficiency based on the discrepancy between the energy that power plants report they produce and the amount of energy sold to end customers.

Experts note that the capacity for generating power from wind in Wyoming and other western states will outstrip the transmission infrastructure if more is not built. Mike Schmidt at LCCC notes that wind potential in Iowa, where he once taught, is much less than in Wyoming. However, the amount of wind energy Iowa actually adds to the mix of power sources going to consumers is much higher than in Wyoming. That's because in Iowa, which is more populous and developed than Wyoming, there is easy access to transmission towers for moving wind-generated electricity. But in Wyoming, Schmidt says, "If we can't move power somewhere to sell it, it isn't going anywhere."

That's why PacifiCorp and other developers endeavor to engineer large-scale transmission projects. Rocky Mountain Power's Gateway West project that is planned to stretch from eastern Wyoming into western Idaho will add roughly 1,200 miles of transmission line. The company's proposed Gateway South project will move power generated in Wyoming into Utah and eventually Nevada. That's depressing news for those who don't want Wyoming to look like a place where we ranch metal instead of cattle. But it's good news for anyone who wants to make use of Wyoming's abundant winds and help keep carbon particles out of the air.

Even though renewable energy such as wind is considered clean and is being clamored for in many locations around the country, its use is not without consequence. As Walt George noted, "Nothing that people do is without some change or effect. Even solar and wind power change the landscape." Wind-farm installation is a major construction project, complete with roads and heavy equipment and dust blowing in the air. These projects kill, displace, or at least alarm, wildlife. They alter the view. They are worthless without adequate transmission towers to deliver the electricity where it is needed, to places where people live in large numbers and demand cheap, plentiful power. Laurie Milford, director of the environmental advocacy group Wyoming Outdoor Council, understands why many of us have a hard time leaping with both feet onto the wind bandwagon. She's an outdoors lover who never looks wind whipped or sunburned, quite a feat in Wyoming. She told me she's made it one of her missions to teach her group's members that wind power isn't without its costs. She uses the sage grouse as an example. The bird, a little larger than a quail, has been in decline due to habitat fragmentation, much of it a result of energy development. If the grouse were to be placed on the endangered species list, so would be their sagebrush-dwelling companions, such as pronghorn antelope, ferruginous hawks, pygmy rabbits, and Wyoming pocket gophers.

"All of this is a dilemma for environmentalists," Milford said.

> We've been calling for decades for our country to develop renewable energy. Now that it is about to happen on a big scale, we cite sage grouse concerns, and threats to bats, raptors, and viewsheds as reasons to slow down. So which is it? Is commercial-scale wind development part of the answer to our power conundrum? Or are wind farms the new bane of Wyoming residents who are tired of giving up our landscapes and our wildlife to export energy elsewhere? In reality, of course, it doesn't have to be an either/or proposition. We're hopeful we can work toward increased energy efficiency as a country, encourage more renewable energy development in this state, and at the same time protect Wyoming's natural resources.

Understanding the complexity of these competing positions is the job of Aaron Clark of the WIA. Wyoming creates much more energy, renewable and otherwise, than its population of just over half a million people could ever use, and much of it is exported elsewhere. Many in this state wonder why we are taking chances with our environment and coping with the upheaval in our communities by shouldering the burdens of energy production. Clark says there could be no meaningful shift from nonrenewable to renewable energy sources in this country without Wyoming. That's one reason the state has put in place strict regulations for permitting. As Clark notes, "We want to be responsive to the energy needs of others but not trash our state."

One person who knows a lot about the drive to develop wind is Grant Stumbough. An outdoorsman with a mustache and white cowboy hat, he is the area coordinator of the Southeastern Wyoming Development Council. Many developers have noticed that Wyoming has great untapped wind potential, much of it located in southeastern Wyoming, an area mostly already developed by ranches or small communities. Stumbough helps landowners understand the potential that developers are busily tapping into their calculators even as they drive through the gate.

"We have a tremendous wind resource here, and developers are recognizing that and knocking on doors," he said. "I was talking to some of these guys and they said, 'These guys are coming and we have no idea what our wind resources are worth.'" So Stumbough brought landowners together to consolidate parcels of land to make it more marketable to developers compared to negotiating with many individual property owners. The idea is to offer what developers want—wind power—and locate it in a place that is appropriate and acceptable to most parties.

To those who object to wind development, Stumbough points out that when ranchers have difficulties meeting payments on loans or property taxes, they sometimes end up selling off parts of their property. And the buyers are often real estate developers who will sell the land in parcels for ranchettes of just a few acres. These ranchettes break up the undeveloped land with roads and fences, which disturb wildlife and are not aesthetically pleasing to people who appreciate Wyoming's open spaces. "Would you rather have wind farms or subdivisions?" he asked.

The developed land that Stumbough is concerned with in southeastern Wyoming sits on abundant wind resources on topography that is appropriate for wind development. It is near existing transmission towers and roads. It has appropriately gentle terrain with windy ridge crests. It isn't surrounded by sensitive environmental areas. Its existing functions, such as cattle ranching, can still be continued even under the shadows of wind turbines. But developing commercial wind farms on land owned by individuals is costly to power companies. Landowners must be paid for the use of their land, sometimes in a lump sum and other times in royalty payments, according to the *Commercial Wind Energy Development Resource Guide* produced by the University of Wyoming.

Wyoming's governor during much of this wind rush, Dave Freudenthal, organized a coalition of state leaders and interested citizens to make sure Wyoming develops its wind potential in a smart way. He has stood with one foot in the stirrup of two runaway horses, bringing together both pro- and antidevelopment people. During a symposium at the University of Wyoming, he recalled a period after the oil embargo of 1973, when Wyoming was poised as a key player in the national energy policy. Wyoming up to that time had mined only a modest amount of coal underground. Then technology arrived to make strip mining of Powder River coal economical. "Suddenly Wyoming was action central for development of coal," he said. Fast forward thirty years. "Now some people come into my office and say it'll be the end of the world if we develop wind." Other visitors to his office are developers who say that if the state isn't "nicer" to them, they'll pull out, he said.

Freudenthal tries to find a middle road between these pro- and antidevelopment positions, saying, "I don't think we ought to be rushed. We'll be at this for several years." He stresses the importance in developing ground rules for development. Wind is the most highly subsidized form of energy production in the country, far larger than oil and gas, or coal, he said.

"We're glad to host the turbines but we're expecting the people who put them up to do a little more. It is not the case that we're going to be happy being some colony with a bunch of towers sticking up in the air, and no jobs," he said.

Even though the governor sees tremendous potential in wind development, he still must make its proponents play by the same rules that guide coal, gas, and oil. For example, they need to be sure their activities do not harm sage grouse. If they do, and the sage grouse becomes listed as endangered, Wyoming's economy and way of life would be seriously disrupted. This was an issue also raised by Molvar and Milford in their descriptions of environmental concerns.

The governor's office says if the sage grouse were to be listed, nearly 80 percent of coal production would be subject to additional reviews; 83 percent of natural gas production, 64 percent of gas production, and almost 40 percent of activity on private land would be reviewed as well. Some folks would see this slowdown of development as a very good thing, indeed. But it also means that if a rancher wanted to move his cattle from one pasture to another, a federal wildlife manager would have to be called in to search for sage grouse breeding areas before the rancher could continue. That would mean delays in even the most mundane of daily ranch operations.

Freudenthal concurs with Mike Schmidt, Laurie Milford, and others that the entire wind conversation in Wyoming would be moot without determining how to transmit the power to the power plants in the right way. The most pro-economic-development people in the state suddenly acquire a bad case of not in my backyard—or NIMBYism—when faced with the prospect of wind farms and transmission towers on their own turf, he said. As a result of these discussions, in 2011 Wyoming passed the Wind Energy Rights Act, which codifies regulations, tax structures, and procedures for developers and landowners interested in hosting wind farms.

It was in the early period of wind development in Wyoming that I took my tour of the Foote Creek Rim facility. On the day Tony Kupilik and I stepped out of the Dodge truck, we stood in a relatively modest breeze at the top of the rim. We gazed out across the wide open basin, our eyes following the power transmission lines to the northwest. We could almost make out the closed coal mines at Hanna, along with the tiny towns of Rock River and Medicine Bow. The ground under our feet was pricked with early June wild flowers, and the first of the Indian paintbrush poked up their red heads to welcome the buzzing bees. A moment later we spotted, literally in the shadow of a spinning wind turbine, an hours-old pronghorn antelope that lay so still I first took the small gray form for a rock. We'd seen a doe antelope trot off a few moments before so we took a few cautious steps for a closer look. She stood staring at us from about one hundred yards away, keeping watch over the scene. The little fawn never moved a muscle nor blinked one of its prominent eyes while we whispered and snapped pictures. We had no doubt the doe would return as soon as we drove off.

Kupilik is a booster of wind energy but doesn't believe it will answer all our nation's energy problems. "Is wind the solution?" he asked rhetorically. "Probably not. Is it part of the solution? Probably so." One problem he sees is the large footprint of a wind farm. He explained that a coal-burning power plant would take up far less space than this facility. The footprint could be smaller if there were taller towers, but then other problems both environmental and aesthetic would need to be solved. He thinks nuclear power is the way the country will need to go.

Eventually Kupilik and I meandered back to the truck, both of us stopping to photograph wildflowers and barely noticing the low swooshing hum of the wind turbines. I took pictures of wind turbines but not of the sensitive cultural artifacts that Kupilik asked me to avoid. We discussed the various types of wind turbines and chatted about "pitch" and "yaw" as he explained the fine points of variable-pitch versus fixed-pitch turbines. Before climbing back in the truck I asked if he ever felt like throwing a tent in its bed and making camp up here some nights. "Yes," he replied. "Really?" I asked? "Yes," he said again. I wasn't sure if he meant yes he thought about it or yes he did it. We both knew he is lucky to find himself working in a place he so treasures as an outdoorsman. "It makes you think you had it planned all along, the way things work out."

2

Angels and Monsters

A Wyoming Coal-Fired Power Plant

FEBRUARY 1

A chilly, windy start to February in Laramie. The low overnight was in the high teens. The National Weather Service predicts a high today in the mid-30s, with average sustained wind speeds around twenty miles per hour and no precipitation. All things considered, not too bad.

Last winter we didn't see the asphalt pavement of our street between the first of December and the first of May. Our short block twists to the northwest as it heads downhill. The low winter sun can't do much to melt the snowpack unless it gets help from the wind and the temperatures. Last year the wind and sun were game but they needed the warmth to show up and help. This year it has, from time to time, and now I can see open black pavement reclaiming the street from the stubborn ice.

This morning we are moving a bit slowly, recovering from a weekend of parties at our house. Two nights ago was our annual Groundhog Day party, which is a good excuse for friends to come out of their own winter burrows and celebrate the spring equinox with us. The patron saint of our party, a now-deceased Wyoming groundhog called Lander Lil, is now just a statue on a street corner in her hometown. Each year a few folks in Lander gamely gather at that corner to observe whether or not her bronze form will cast a shadow on the concrete beneath her feet. Like Lil, her more famous counterpart, Punxsutawney Phil, also saw his shadow, for a sure six more weeks of winter.

Last night we polished off party leftovers during a small Super Bowl gathering at our house. Phil's state-mates, the Pittsburgh Steelers, prevailed for the sixth time, this time over the Arizona Cardinals. Pittsburgh, in western Pennsylvania, sits on the largest anthracite coal deposit in the country. It is a coal-mining, steel-producing, winter-coat-wearing, football-gaming town. The Cardinals play in Glendale, Arizona, outside of the desert landscape of Phoenix, which gets lots of its electricity, in one way or another, by using precious water in service of one of the fastest growing metropolitan areas in the United States.

After two parties I'm in the mood for some quiet time, so I'm reading Elizabeth Kolbert's *Field Notes from a Catastrophe*, which chronicles human-caused global warming. I'm trying to understand the difference between naturally occurring greenhouse gases that keep our planet's climate warm enough to be habitable, and the human-caused emissions of carbon dioxide that put the planet in jeopardy of a meltdown. I'm particularly mindful of these issues at this time of year, because this is when we have to make a decision about our gas utility service. Each year, we have to decide, with basically no data to help us, among several gas service providers and whether we want our bill to rise and fall with market prices or if we should take the "pass-through" rate, paying just what the gas company pays plus a specified markup. Not being experts in commodities futures, we usually just follow the advice of a Wyoming journalist friend who has covered the story. "Take the pass-through rate. Anything else will just get you screwed," he told us.

Kolbert's book directs me to the Environmental Protection Agency's website, where I calculate the carbon footprint for my household. I plug in the various factors such as the amount of electricity we use, as stated on our bill, what sort of car we drive and how far we drive it, and whether and what we recycle. We have two vehicles, one a midsize SUV that seems a necessity for our quasi-rural western lifestyle. But admittedly, we also drive it for puttering around town after groceries and such. The good news is that our town covers only eleven square miles so we don't go very far. And in spite of having a full-time home office, we still use far less electricity than the average American household. The final power saver is that at 7,220 feet of elevation, there's no need in Laramie for air conditioning. According to the carbon footprint calculator, the average U.S. household of two contributes a yearly average of 41,500 pounds of carbon dioxide into the atmosphere. I am happy to learn our household was responsible for releasing a modest 32,371 pounds of carbon dioxide per year.

It is hard to feel smug, though, after reading Kolbert's book, which ends by inviting readers to ponder whether humankind will find a solution to climate change before it is too late and the current patterns of global warming create irreparable change to life on earth. Kolbert's final words suggest that we are showing no signs of using our advanced technologies to save ourselves, but are instead self-destructing.

Since the days of the last energy crisis, during the presidencies of Gerald Ford and Jimmy Carter, I've been accustomed to wearing lots of sweaters in the house rather than kicking up the heat. My father was fond of saying, "When not in use, turn off the juice," and flicking off light switches in rooms we'd vacated even if just for a few moments. Thus, as a survivor of the 1970s gasoline shortages and a captive of the wallet-squeezing price spikes of the recent years, I've become accustomed to combining errands and walking or biking rather than driving, when I can. But now I consider whether I need to use appliances during peak times. I wash most of my clothes in cold water. I limit the use of television and radio merely as providers of background noise. These steps are tiny compared to the changes we'd all need to make in our behavior if we eliminated fossil-fuel power plants, specifically coal-burning ones, before renewable energy can fully take its place.

The Energy Information Association (EIA) has an excellent website and I find myself lingering over its Basic Information pages, which are clearly written and address issues of interest to novices of all ages. From those pages, and the website in general, I learned that coal provides about half of the electric power in the United States. Coal is not the only fossil fuel burned in electric plants—some plants are oil-fired or burn natural gas. Together these fossil fuels provide about two-thirds of the electricity in this country.

Many believe that burning coal for electricity is akin to driving a stake into our planet's heart. Burning coal fouls air, spoils water, and puts big gouges into or onto the surface of the earth. Underground miners get lung diseases. Trains that drag coal from the mines to the plants offload soot and smog. Yes, burning coal makes a large mess that even the power plants and utility companies must fight against each day, whether they want to or not.

Burning coal does release gases into the atmosphere: that is not a secret. Among those gases is carbon dioxide, the main greenhouse gas linked with global warming. Burning coal also releases sulfur, nitrogen oxide, and mercury, polluting the air and water, according to the EIA. Sulfur mixes with oxygen to form sulfur dioxide, a chemical that can affect trees and water when it combines with moisture to produce acid rain. Emissions of nitrogen oxide help create smog, and also contribute to acid rain. Mercury that is released into the air eventually settles in water. The mercury in the water can build up in fish and shellfish, and can be harmful to animals and people who eat them.

Coal burned for fuel contributes 2.1 billion metric tons of carbon dioxide annually, or 36 percent of atmospheric particle pollution, according to the EIA. Coal's contribution comes in second to the contribution of burning petroleum for transportation fuel, which contributes 2.6 billion metric tons annually, or 44 percent of particle pollution. The other 20 percent comes from natural gas emissions.

But burning fossil fuels is not going to stop tomorrow. Ken Salazar, secretary of the interior in the Obama administration, said to those gathered at the 2009 Great Plains Energy Expo and Showcase in Bismarck, North Dakota, "I don't want you to be scared, those of you here who are supportive of coal and oil and natural gas. The fact remains that oil, gas, and coal are a very important part of our energy portfolio and will remain a part of the energy portfolio in the future."

The Sierra Club has taken a strong stance against coal and other fossil fuels and toward an ambitious plan for renewable energy sources to take their place. They would like to see 20 percent of our nation's energy come from renewables by 2020. In addition, they scoff at the viability of the "cleaner coal" alternative, in spite of the industry's use of such terminology. They are pessimistic about the economic or practical viability of any of the attempts to make coal cleaner, at least at the present time. However, they don't expect the nation to be able to wean itself off coal immediately, suggesting that some transition time to alternative fuels will be required.

Many environmental organizations agree with the Sierra Club's assessment but show less patience. For example, a group unambiguously called No Coal advocates the immediate shutdown of existing coal plants, effectively going cold turkey on coal. And the League of Women Voters is campaigning for a moratorium against any new coal plants for the next ten years. They cite human-caused global warming and its connection to rising sea levels and unusual weather extremes as the reason. They believe that with better conservation and efficiency in the things we power with electricity, we won't need to burn any more coal.

Many metropolitan areas are headed that direction. For instance, the Colorado Public Utilities Commission hopes to eliminate coal plants in the Denver area by 2017 to meet Colorado's clean-air standards. Specifically, five coal-fired power plants in the Denver area will close, a new natural gas–fired plant will be built, and two coal plants will be transitioned to natural gas fuel.

It is often less expensive to mine for surface coal than it is to drill for natural gas. That's one reason that when the price of natural gas skyrockets, developers are encouraged to propose new coal plants to take advantage of that less expensive fuel source. The Department of Energy's (DOE) National Energy Technology Laboratory tracks the progress of new power plant development. According to their 2011 report, "Experience has shown that public announcements of power plant developments do not provide an accurate representation of eventually commissioned power plants. Actual plant capacity commissioned has historically been significantly less than new capacity announced." But as of July 2011, eleven new power plants were under construction and one near construction, for a combined capacity of 7,704 megawatts from burning coal. The year 2010 was active for new coal plants, too, after many years of very few coal plants coming online. In fact, the DOE reports that more new capacity came online in 2010 than in any of the previous twenty-five years.

Because renewable energy has gained some political momentum in the last several years, many announced coal plants have been tabled or at least seriously delayed. The Sierra Club maintains a useful database of proposed new coal plants whose status they describe as either active, upcoming, uncertain, or progressing. Another category that they label "victory" indicates that development of the plant has been stopped.

The irony is that these newer plants could have replaced some of the nation's fleet of highly polluting older plants. Many of these plants are so old they were exempted from regulations of the Clean Air Act, first passed in 1963, limiting sulfur dioxide emissions, which contribute to global warming. Updating their pollution control systems would have cost too much. Closing them would have deprived their customers of electricity. Instead, they continue to operate and pollute. No one wants to go back to the days of rolling brown outs, it seems.

A proposed new coal-fired power plant in Kansas, known as the Sunflower Plant, has been engaged in the permitting process for several years. The Kansas Department of Health and Environment denied the building permit for the 895-megawatt plant in 2007, but in late 2010 moved to allow the plant to be built. Karen Dillon of the *Kansas City Star* writes that the go-ahead permit was a "huge victory for Sunflower Electric Energy Corporation because the announcement comes just days before Jan. 2 [2011] when new federal regulations would have required expensive greenhouse gas controls be installed."

The environmental organization Earthjustice is opposed to this and other coal-fired power plants. Their take is that the final permit was pushed through by the powerful coal industry "despite a comment period that generated 6,000 public comments, many of which were against the project, despite decreasing electricity demand, low natural gas prices, and considerable renewable energy growth." They note the irony in the fact that power will be going not to rural Kansans, but to city dwellers in the Denver metropolitan area, the very place that is eliminating its own coal-fired plants.

Stories like these of coal plants in Colorado and Kansas are being played out in similar ways all over the country as state governments face or impose their own tightening regulations around clean air and old power plants. The oldest coal-fired plant in the country is the Perry K. Steam Plant in Indianapolis, Indiana. It began operation in 1938, just a few months before the next oldest plant, the Blount Street Station near Madison, Wisconsin. To put the distance of that day into context, in that same year, Orson Welles broadcast his radio version of "War of the Worlds," sending people into a panic over a fictional invasion from Mars. Another thirty-three plants operating in the United States were built in the 1940s, the decade dominated by World War II and in which only 55 percent of U.S. households had indoor plumbing.

The dirtiest coal plants in operation tend to be the oldest. Plant "dirtiness" can be measured by carbon dioxide emissions or emissions of substances such as sulfur dioxide or nitrogen oxides. According to the Sierra Club chart, the most polluting plant is the R. Gallagher Generating Station near New Albany, Indiana. It was completed in 1961 and is operated by Duke Energy. It is a 600-megawatt-capacity plant, which emits 50,819 tons of sulfur dioxide each year. The Mojave Generating Station, owned by Peabody was formerly the dirtiest plant in the western United States, emitting up to forty thousand tons of sulfur dioxide annually. It closed after Navajo and Hopi tribes ended Peabody's use of the Black Mesa aquifer. Other shutdowns in recent years have included plants in Tennessee, North Carolina, and Pennsylvania. Sometimes plants close because they are simply slated to be retired once they reach their life expectancy. Others close because upgrading their pollution controls would be too costly. For example, in 2010, the Oregon Environmental Quality Commission approved Portland General Electric's plan to close the state's only remaining coal-fired plant. The plant, in Boardman, Oregon, began operating in 1977 and is the youngest plant in the nation to be closed for environmental reasons. It was scheduled to remain open until 2040 but would have needed more than five hundred million dollars of pollution-control equipment on the plant by 2017 in order to comply with federal and state clean-air standards, according to Portland General Electric, which operates the plant. Now PGE will install only about sixty to ninety million dollars' worth of pollution controls, allowing the plant to operate within legal parameters for another ten years. PGE said the Boardman Coal Plant is a prime source of cheap, reliable power and that closing it will drive up utility rates. PGE will also have to find ways to replace the power created at this 550-megawatt-capacity plant. And the 130 employees in Boardman, Oregon, will need to find other jobs.

Boardman emits about four million tons of greenhouse gases a year and another 25,500 tons of other pollutants, primarily sulfur dioxide and nitrogen oxides. The Sierra Club, which fought for the plant to close, is suing PGE for not installing enough pollution-control devices when the plant opened. They hope to bring about the closure of the plant as early as 2016.

Power companies describe coal plants as clean and efficient if they use low-sulfur coal and invest in devices to capture the released carbon. The Southwestern Electric Power Company (SWEPCO) describes how its John W. Turk Jr. Plant, under construction in Arkansas, meets "emission limits that are among the most stringent ever required for a pulverized coal unit. The plant's 'ultra-supercritical' advanced coal combustion technology will use less coal and produce fewer emissions, including carbon dioxide, than traditional pulverized coal plants."

In spite of these technical advancements, the plant has faced legal opposition from the Sierra Club and others over air- and water-quality issues. SWEPCO has settled some of these lawsuits in court but challenges from some organizations continue. According to SWEPCO, "Two litigants—the Sierra Club and the National Audubon Society/Audubon Arkansas—are continuing to challenge the air permit before the Arkansas Court of Appeals and the Corps' permit in a companion case still pending before the U.S. District Court in Texarkana. SWEPCO will continue to aggressively defend the permits issued for the plant."

The Boardman plant is alive, although in many ways it is already dead. The Turk plant is still struggling to get on its feet. Sometimes it seems coal has become a zombie fuel, politically dead in its present form but still walking among us, wreaking havoc. Most states have a nonbinding standard for their renewable energy portfolios, but California's is the most ambitious. It is restricting the use of coal in its power plants as it seeks to replace fossil fuels with renewables on a very ambitious timetable. As Sharon Beder notes in *Power Play: The Fight to Control the World's Electricity*, regulatory events that happen in that large market tend to trickle out to other states. Their present goal is to reach 33 percent renewable fuels used by state utilities by 2020. That makes coal-producing states like Wyoming nervous about whether their product will find a market in power plants in California, the eighth-largest economy in the world.

Even if as a nation we decide never to pulverize and burn another ton of coal, the fact is, it is a plentiful fuel source. In fact, the United States has the world's largest known coal reserves, enough by most estimates to last approximately 225 years at today's level of use. About 92 percent of the coal used in the United States is for generating electricity. It powers our homes and our industries. We wouldn't have steel, or clothing, or automobiles, or wind turbines, without it. We also wouldn't have huge denuding strip mines in the West, or chunks of mountaintops blown off to loosen coal in the Appalachians.

The biggest supply of coal in the United States is in the Powder River Basin of northeastern Wyoming. In this region, grass-covered basins sprawl as far as the eye can see. Most people around here ranch and also find work in the gas fields and coal mines. The largest town is Gillette, population around twenty-nine thousand. Its residents tend to be younger and more representative of the Y chromosome than most of Wyoming. Their annual income exceeds that of the rest of the state, too. The reason: the extractive industries pay pretty darn well, and folks capable and willing to do hard, noisy, dirty work—usually men—are worth their weight in coal.

I wanted to pay a visit to the coal-mining part of the state, about three hundred miles away from Laramie, in order to tour one of the several coal mines in the basin. A few of the mines operate summer tours for the public. I wanted to take a solo tour as a journalist, as I had done at the Wyoming wind farm. In my attempt to do so I racked up dozens of unanswered e-mails, unreturned phone calls, and two appointments rescheduled then dropped completely. I had hoped the Arch Coal mining-company representative based in Golden, Colorado, would get me into their mine, where I could then make contact with the folks doing the actual mining. Apparently he didn't relish the idea of making a five-hundred-mile round trip for an afternoon at a mine and eventually stopped returning my calls. I decided that if I couldn't get the sort of insider view I wanted, I'd talk with someone who had. That brought me to John McPhee: journalist, teacher, *New Yorker* contributor, and author of nearly thirty books, some of which are about geology and the West. By way of full disclosure, John McPhee has long been my journalism hero, but I'd never met him or spoken with him until I whimsically contacted his publisher to request an interview. I had no expectation I would hear back from him, since he's renowned for giving few interviews. Imagine my surprise when a few days later my phone rang and a voice as grave as a preacher said, "This is John McPhee calling from Princeton, New Jersey." I tried to sound cool, like I was accustomed to talking to lions of literature every day. But there he was on the phone taking my questions about how to bust down a corporate wall more daunting than the tall cyclone fencing around a coal mine.

John McPhee wrote a long article called "Coal Train" for the *New Yorker*, which reappeared in his 2006 collection *Uncommon Carriers*. In it he describes his sojourn on a coal train's regular run from Marysville, Kansas, to North Platte, Nebraska. His companions, engineer Scott Davis and conductor Paul Fitzpatrick, were on their leg of the journey to collect coal from the Powder River Basin mines and haul it to its eventual destination at coal-fired power plants to the east.

McPhee affirmed my convictions about the importance of

seeing firsthand the places one wishes to study. “There is no substitute for being there, for describing the height, depth, and range of a place” he told me. He also affirmed my recent lesson about large corporations being, in his words “not cooperative” and even “rejective” when it comes to communicating with journalists. He suggested that I should just drive up to the Powder River Basin, follow the maze of roadways, and peer where I could through the fence.

It occurred to me that following McPhee’s approach, I already had seen the mines. I’d spent more than a little time in the Powder River Basin with my husband, conducting research for a book on small-town Wyoming bars. We’d seen the railroad-man camp at the town of Bill, population one, just south of the mines on Wyoming 59. That’s where railroad workers stay between shifts of hauling out coal. My husband and I had gazed into those monstrous holes in the ground many times along Interstate 90 and various back roads. Even though I hadn’t been hard-hatted and guided around the terraced gorges, I know what those Wyoming strip mines look like and sound like, and that being in their midst makes me mourn for the chunks of Wyoming taken away forever.

The total amount of coal in Wyoming is more than 1.4 trillion tons, according to the Wyoming Mining Association. They estimate that Wyoming’s recoverable reserves, in seams thicker than five feet and buried less than one thousand feet deep, total more than forty billion tons. Much of it is in the Powder River Basin, which splays over five of Wyoming’s twenty-three counties. To get the coal out of the ground, a mining company first explores the properties of a suspected coal seam. They drill a hole three to six inches wide and three hundred feet deep through the overburden, what most of us would call “the ground.” They bring pieces of rock up from that deep hole for geologists to examine.

If they determine the coal is appropriate for mining, it is time to scrape away the materials of the ground that are not coal. To do that they use a house-sized excavation machine called a dragline. To picture the scale of mining equipment, imagine for a moment King Kong with tiny Fay Wray grasped in his fist. The dragline is powered by electricity and is basically connected to an outlet by a really long extension cord. The overburden is heaped in a big pile then scuttled away in bulldozers and haul trucks.

After the overburden is cleared, the ground is terraced into a series of benches. The seam is divested of its coal by being blasted with up to sixty thousand pounds of ammonium nitrate fertilizer at a whack. After an earth-shaking *ka-bamm*, the loosened coal is loaded into trucks. Again, think of King Kong and what he might drive, and you'll get a sense of the truck size involved. Each truck can haul one hundred tons of coal. The trucks haul the coal to a conveyer belt at the structure known as the tipple. From there the coal is loaded onto the awaiting trains, which are typically more than a mile long. The trains average 130 cars each. Seventy to eighty trains leave the basin every day. Some of them are eventually driven by John McPhee's Union Pacific companions.

After the coal seam from a given pit has been exhausted, all that overburden that was removed once again has a role in the attempted normalizing of the area. It is returned to the pit, where reclamation crews labor to put the place back together again, a little like the repair of the Scarecrow in *The Wizard of Oz*. They get a little bit of rock from here, a little bit of sagebrush from there, and before you know it, they've cobbled together a landscape. Of course, some transformations can never be disguised, like the roads that range around the various mines and the train tracks constructed to haul the coal away.

The Bureau of Land Management, which manages this basin land, says that almost 14 percent of U.S. carbon dioxide emissions originate from coal that is mined in the Powder River Basin. Seeing that statistic and then pointing a finger at Wyoming for being a colossal industrial polluter is to be both accurate and a bit unfair, because about half of the electricity produced in the United States comes from the Powder River Basin, too. Powder River Basin coal is low in sulfur, giving it an advantage over its higher sulfur, more polluting cousins. Coal formed in Wyoming is the sub-bituminous variety and has taken about a hundred million years to form. It does not burn as hot as the bituminous, higher-sulfur coal of the underground coal mining areas of the Appalachians. Bituminous is the most abundant type of coal in the United States, and is between one hundred and three hundred million years old. The anthracite coal in Pennsylvania is a very hard coal that burns extremely hot. It is used mostly for heating homes, not for commercial production of electricity. Another form, called lignite, is the youngest of coals. Lignite occurs in many places but is mined mostly in Texas.

As our national energy policy and our household economies trend away from fossil fuels and toward "going green," the Obama administration promises to ramp up the contribution of renewables such as wind and solar power to our nation's energy portfolio. In fact, the Obama administration's energy plan target is that 10 percent of our electricity will come from renewable sources by 2012, and 25 percent by 2025. Currently, about 8 percent of our energy comes from renewable sources, including hydroelectric, biomass, wind, geothermal, and solar. But it will take time, technology, and capital to reach President Obama's goals. So in the meantime, few observers think coal-fired electric power plants are headed back to the days of the dinosaur, from whence the fossil fuels they burn came. There are still about five hundred power plants, with their smoke-stacks and coal conveyors, operating in the United States. They will continue to populate our environmental landscape for years to come.

One of those plants is the Laramie River Electric Station near Wheatland, Wyoming, about 80 miles from where I live. It is one of nine coal-fired power plants in Wyoming and is about

150 miles south of the Powder River Basin, where it gets much of its coal. I wanted to find out how the plant produces the electricity I use every day and what steps it takes to mitigate environmental damage. I felt deeply conflicted by the argument that pollution is a short-term necessity, knowing that some of the damage caused by burning coal could linger for generations. I felt John McPhee sitting like an angel on my shoulder repeating his advice and urging me on: there's no substitute for seeing a place and absorbing its size and scale to understand what it is all about. So get in your car and drive up there.

Unlike my thwarted trip to the coal mines, I had no trouble setting up a solo tour of the power plant. I arranged a tour for one July afternoon when the temperature in Wheatland was a characteristic 95 degrees. Richard Bower, who among other tasks gives approximately seventy tours of the station each year, arranged to meet me at 1300 hours, his use of military time common at power plants. He's comfortable with it, possibly a holdover from his days in the Marine Corps during the Vietnam War. I pulled up to the gate, picked up the phone in a box, and announced myself. The chain-link gate pivoted open to admit me and in I drove, parking in front of the visitor center. Its lobby, all-glass walls and rich wood interior, seemed more like a fancy hotel than a power station alone on the high plains of eastern Wyoming.

Bower greeted me right on time. With his straight posture, cool blue eyes, deep husky voice, crisp dress shirt, and tan slacks, he reminded me of a better-looking Lee Marvin. His Looney Toons–character tie was only a bit incongruous. He handed me a hard hat and ear plugs, and with the assistance of the office ladies, helped me decide which style of safety glasses would best fit my face.

The official tour started in front of a bank of illuminated maps, diagrams, and various displays about how coal comes from the ground, gets transported to the plant, is pulverized and burned to boil water, to create steam that turns turbine blades, that spin generators, that create electricity, which zaps through transmission lines and eventually powers our homes and industries. Bower explained that the Missouri Basin Power Project, a group of six electric utilities, owns the plant and adjacent facilities. North Dakota–based Basin Electric Power Cooperative is a 42.27 percent owner of the project and is also its operating agent.

Three coal-based units comprise the Laramie River Station, each with 550 megawatts of capacity. Units 1, 2, and 3 began operating in 1980, 1981, and 1982, respectively. Laramie River Station is unique because it delivers electricity to two separate electrical grids. Unit 1 is connected to the Eastern Interconnection, while units 2 and 3 are connected to the Western Interconnection. The electricity produced at Laramie River Station is sent to substations in Wyoming, Nebraska, and Colorado.

After this lobby-map orientation session, it was time to don the safety gear and start the tour of the plant proper. Bower rattled off statistics about the plant's construction as we walked. Boiler height: 225 feet. Stack height: 605 feet. Construction costs: $1.6 billion. Environmental controls investment: $330 million. Horsepower: 600,000 per each of the three units. Fuel consumption: 375 tons of pulverized coal per hour per unit.

Considering that the plant employs roughly three hundred people, we saw hardly a soul on our tour. Perhaps most people don't chose to spend 95-degree days traipsing around a power plant, taking in sights like the electrostatic precipitator, the sulfur dioxide scrubber, and the coal silo. Between the heat and the ever-present sound of screeching metal, I quickly realized that people who work in this industry are a heckuva lot tougher than I am.

Our first stop on the tour was the seven pulverizers, which reduce coal to dust for burning. My vision of the plant burning coal the way I burn briquettes in my Weber grill flew from my head. The stuff that is burned isn't a lump, it is a powder.

From the pulverizers, we walked by the bottom-ash hoppers. When coal is burned it leaves behind a heavy gray mess called bottom ash, which is collected in a hopper. The bottom ash is periodically removed by high-pressure water jets. The resulting slurry can be recycled as roadway material or for other uses. The rest of the bottom ash goes up the flue and is captured and recovered as fly ash, which can be recycled for use in concrete. Ash is typically stored in a retention pond. This works pretty well except for when it doesn't.

On December 22, 2008, the dike around a pond that had been built to retain fly ash at the Kingston coal plant in Harriman, Tennessee, collapsed. One billion gallons of sodden ash flowed like cold lava covering three hundred acres around the plant. According to the Environmental Protection Agency (EPA) and local reports, the result was 5.4 million cubic yards of sludge up to six feet deep that destroyed homes and property, washed out roads, ruptured a gas line, broke a water main, obstructed a railroad track, and destroyed power lines. In short, this was the country's biggest industrial disaster and has prompted skepticism about the notion that coal could ever be "clean."

There is a retention pond at Laramie River Station, and much of the ash accumulated there is used in the roadways on the property. There are several wastewater holding ponds at the site in addition to the pond for ash. There are also ponds for scrubber waste, sewage treatment effluent and sludge, water-treatment-plant waste, and plant site runoff. As Bower drove me in the power plant van, a few ducks that normally frequent the nearby reservoir were paddling in one of the ponds. I wondered if they'd be harbingers of ill effects. But all of these ponds, and everything else about the plant's potential to pollute, is closely monitored by the Wyoming Department of Environmental Quality (DEQ).

The Wyoming DEQ was created by the Wyoming Legislature in 1973 after passage of the federal Environmental Quality

Act. It is a busy place. The agency sends out numerous notices of violations each month to businesses and individuals that violate regulations designed to protect air, water, and land quality. The agency posts on its website these notices of violation and whether or not the violations have been resolved. Violations range from innocuous-sounding omissions, like failure to deliver an annual report, to the more serious, like failure to control fugitive dust (at a coal mine); incorrect discharge of effluent (in a town); and illegal disposal of used oil (at a transportation business). It is both easy and disconcerting to imagine the state of the environment without government agencies like the DEQ inspecting facilities and without citizens making complaints about coal dust in the air or oil on a local fishing pond. A review of the DEQ site didn't reveal any violations by the Laramie River Station. Rather, it reinforced Bower's assertion that the DEQ and the power plant strive to ensure that the plant remains in compliance with environmental and other regulations, avoiding violations and costly fines.

Not that this task is an easy one. The DEQ has identified many potentially polluting components at the Laramie River Electric Station. The three pulverized coal-fired boilers; the various coal, limestone, and ash handling facilities; the coal and limestone storage piles; the various fuel and lube oil storage tanks; and miscellaneous small pieces of diesel-fired equipment, are all under the scrutiny of the DEQ and a dozen other regulatory bodies. The plant regularly upgrades pollution-control devices to limit haze and emissions, which costs millions a year, but beats the alternative of being fined or, worse for them, shuttered.

Richard Bower and I left the ash ponds behind and took an elevator to the third floor to the turbine-generator deck. This deck was at once one of the hottest places on the tour and the coolest in terms of the I-can't-believe-I'm-standing-here factor. With the temperature reading over 100 degrees, Bower at last broke into a controlled sweat as he hollered over the thrum of the turbine generator about how electricity is created. My point-and-shoot camera and digital voice recorder were as useless at absorbing and documenting this information as were my safety-protection muffled ears. The seven-inch-thick floor decking vibrated: I was standing on the thunderbolt forge of Zeus and I couldn't hear a thing.

Just when I feared Bower was going to open another door and coax me deeper into the bowels of the plant, he led me to the computerized control room that runs the entire operation. There I found blessed air conditioning and near silence. I also found Tony Francis, Nick Cancino, and Joe Holtzclaw hard at the sort of work that appears at first glance not to be work at all. Francis gave me the basics on what I was seeing on the dozens of quietly humming computer monitors. Each displayed schematics of the various plant systems. If something went awry, one of the operators in the control room could pick up a phone and dispatch a less-fortunate plant employee to confront the balky machinery and fix it.

After the noisy and hot plant, the control room was so pleasant that Bower and I were reluctant to leave and instead hung around to chat up the control-room operators. Soon the talk turned to speculation about the future of coal-fired power plants. Francis said he thought that plants like this one would always be a piece of the puzzle, meeting the country's energy needs. He said he thought nuclear, wind, and solar power would also fill a role. Bower added that the Arctic National Wildlife Refuge (ANWR) area in Alaska should be drilled. "What's up there?" he asked with his best Lee Marvin bombast. "Caribou?" But Cancino said he thought that the caribou should be protected along with the other natural features that make ANWR irreplaceable.

After about half an hour, Bower and I reluctantly said goodbye to the fellows sitting cool in the control room and headed up to the hot roof of the boiler room. From there we could see much of Platte County's farmland and Gray Rocks Reservoir, which the power company created by damming the Laramie River. One local benefit of damming the river is that locals can spend hot summer days boating and pulling fish out of the eight-mile-long reservoir, whose cool waters were tempting me from the scorching roof. The reservoir normally provides an adequate water source for the plant. But during times of drought, like when I visited, the plant doesn't draw down water from the reservoir to use to create steam. Instead, the power company leases fifty or more wells from ranchers in the area. Since the typical roof of a powerhouse on a warm summer day is 120 degrees, I was grateful when the breeze showed up.

In addition to seeing the understated topography of Platte County, we also had a good view from the roof of some of the plant's environmental control equipment. Bower explained that when gas leaves the steam generator, or boiler, it enters the electrostatic precipitator. That's where about 99 percent of the fly ash is electrostatically charged and collected on a series of large collector plates. When the plates are mechanically rapped, the ash falls, gets collected in hoppers and conveyed to temporary storage silos. The flue gas, sans fly ash, enters the gas desulfurization system, also known as the sulfur dioxide scrubber. It is sprayed with a limestone slurry to remove the sulfur dioxide to comply with EPA and other governmental regulations. The cleaned flue gas is discharged through the chimney stack. Some plants that burn low-sulfur coal can meet environmental requirements without adding a scrubber. In spite of that, Basin Electric opted to install the scrubber at Laramie River Station.

As Bower was explaining these details to me from the rooftop that was even hotter than the turbine-generator deck, my eye wandered to what I most wanted to see since speaking with John McPhee: the coal trains. He had ridden an empty train from Kansas to North Platte, Nebraska, but with five locomotives and 133 aluminum hopper cars stretching a mile and a half, it still weighed almost three thousand tons. He eventually made it to Bill, Wyoming, aboard an ordinary Chevy Suburban.

However, he and Scott Davis were able to ride along with Davis's Union Pacific Rail Road acquaintances into the Black Thunder mine in the Powder River Basin. McPhee admitted they hadn't asked the mining company, directly. Asking forgiveness is more expedient than asking permission, it seems.

Since I had such difficulty getting into a coal mine, I had no hope of riding on a coal train. But here was my chance to see for myself how they got all that coal out of the hopper cars and into the pulverizer. After I dropped a few more hints, Richard Bower loaded me back into the tour van and drove me past the coal silo, coal stockpile, and finally, to the unloading building.

John Dietrick was at the controls of the rotary dumper, which works like a large clamp. Maneuvering the fingers of the rotary dumper, he grasped a car and slowly turned it upside down and in a black avalanche, dumped the coal into awaiting hoppers and feeders. Like a kid at a circus, I observed the proceedings from a catwalk directly above the action, while Bower stood coolly in the operator's glass-and-steel rotary dumper cab. After the dumping of one car, which took about two minutes, the coal dust lay thick on my jeans and shoes like black dryer lint.

Bower returned me to the main building at 1600 hours at the conclusion of our tour, after the office staff had left for the day. He wasn't going to let me leave without tangible memories of my tour. He led me to a storage space behind a conference room and rooted around in some file cabinets for informational brochures and a Laramie River Station retractable ballpoint pen. As I stood there feeling hot and grubby, covered in coal dust, sensing I had a serious case of hard-hat hair, I noticed something remarkable about my guide in his immaculate shirt, slacks, and dress shoes. "Why aren't you dirty?" I asked, practically stomping my little coal-covered foot. "I never get dirty," he said, standing even straighter. "Most people around here just don't know how to walk."

Most people also don't know how we'll replace coal if we wanted to stop burning it tomorrow. There are many excellent environmental reasons for doing so; however, the specter of what would happen to individual jobs and the national economy, which is tied up in electricity-hungry manufacturing and essential agriculture, is also startling, just for different reasons. On the opposite side of the spectrum from the No Coal crowd are those who note that since we've only got an estimated 225 years left for burning coal if we continue at our present rate, we might as well get on with it while working on a variety of strategies to replace it. Neither of these points of view feels either rational or realistic, after observing the complex infrastructure of coal mining, regulation, transporting, burning, and power creation.

Efforts are underway by geologists and other scientists to discover techniques for using coal in ways consumers can feel better about, whether they consider themselves green-minded or not. At present these technologies each have detractors and supporters. Some researchers are working to turn coal into a liquid as an alternative to oil, according to the World Coal Association. Others are figuring out how to turn coal into a gas for use in electrical power plants. Gasification would mean shipping fuel through a pipeline rather than in the bed of a coal train. Others are experimenting with ways to take existing coal-plant emissions and store them underground, a process known as carbon sequestration. Capturing carbon from the stacks of a coal plant and pumping it into caverns underground is seen by some as sweeping a problem under the rug, by others as a tidy way to keep pollutants out of the atmosphere. In addition, some of that carbon pumped underground might be used to force up reluctant pools of oil and natural gas. Like any industrial process, these require water and electrical power, and in the early stages of research and development are very expensive.

The environmental organization Earthjustice argues that

all of this research and development money, time, and energy would be better spent finding a way to make renewables more efficient and viable on a large commercial scale. Earthjustice staff attorney Abigail Dillen says that her organization doesn't wish to fight against coal plants only to see them replaced with natural gas. She notes that although in some ways a modern and well-run natural gas plant is less objectionable than a coal plant, "natural gas is not a silver bullet." Earthjustice argues for a long-term energy future that relies on neither coal nor natural gas. To this end they are fighting vigorous battles against natural gas exploration and drilling in the Marcellus Shale formation in New York, in the Arctic regions, and in parts of the West, to name a few. "We need power—we can't oppose every project," she said. "But we should be focused on renewable energy instead of just moving from one fossil fuel to another."

Other strategies for mitigating fossil-fuel pollution include increasing the efficiency of renewable sources of energy, such as wind and solar. Paul Wolff of Norris, Tennessee, is a consultant in the coal power industry. He believes that renewable energy is "our only path forward." He started in the power industry in 1986 working for the Tennessee Valley Authority. He has since earned a master's and doctoral degrees in mechanical engineering from Georgia Tech and spends most of his time now analyzing data for the electric power industry, many of which help fossil and hydroelectric power plants improve their efficiency. Wolff thinks the public doesn't understand a very basic law of scale: an average-sized wind farm will not replace an average-sized coal-fired power plant.

"It's a huge technical undertaking to transform our energy generation and way of life, even to life without coal-fired power," he told me. "I agree we have to do it, but we need to approach it in a rational way. How many people who want all coal plants to be shut down have invested serious money in energy efficiency improvements in their house, take cold showers, don't use their AC, or whatever it takes?"

Wolff is succinct in his assessment of our various fuel options. "Coal has two hundred years left, there's a finite supply of fossil fuels, oil has a few decades left, and natural gas has so much boom-and-bust it is not a long-term solution, and it is finite. So we do have to make a transition to renewable energy. Solar energy is beaming down on us, but how much we can capture and covert into consumable energy via solar, wind, and biofuels is a challenge. Coal is just a bridge to get us to a renewable power future," he said.

There's good news and bad news about our coal reserves. The good news is that we have a couple hundred years of it left. The bad news is . . . well, you get it. The zombie fuel still has some life left in it.

Wyoming's governor, Dave Freudenthal, expressed himself on this issue with matter-of-fact pragmatism at the dedication of a new coal-fired power plant, near Gillette, which had been years in the making.

"How is the economy going to move forward in what is going to be a carbon-constrained environment? People say, 'Well let's stop all the coal plants in this country.' Interesting thought, except for the fact that there is no regulatory basis to do it. And more importantly there is no practical basis to do it. This country is dependent on the availability of electricity. We will arrive over the next decade at a new set of standards with regard to the management of carbon in some form and the utilities will comply with that."

The problem of containing the contaminants of coal is not a new one. As early as the year 1285, a committee was formed in London to look into the serious problem of pollution from burning coal. King Edward I enacted a ban on burning coal that was eventually declared law. That lasted until cheap coal was such an attractive option compared to expensive wood that the law had to be repealed. But now, as we head toward the end of our planet's coal reserves, some people are noticing that we might actually have to do something to replace it. One can't just go out and pick up a new commercially viable fuel source. Unlike milk or cereal, power doesn't just come from the grocery store.

Richard Bower at Laramie River Electric Station stays clean while working in a coal-fired power plant because, as he says, he knows how to walk. That knowledge will come in handy as he and the coal industry walk a line between the passions and the conflicts of consumers and politicians in the decades to come. Perhaps the coal industry can join forces with scientists and environmentalists to disprove Elizabeth Kolbert's predictions in *Field Notes from a Catastrophe*. Perhaps we can leave some coal in the ground where it belongs, instead of allowing everything to go up in smoke.

3

Fission and Fishing

A Nebraska Nuclear Power Plant

MAY 1

The University of Wyoming is in its final week of classes for the spring semester. A big discussion in our local newspaper, the *Daily Boomerang*, has to do with the annual "Tour de Laramie." That's the occasion when graduating seniors ride their bikes from bar to bar in downtown Laramie for libations. In their view, bicycling to the bars prevents them from driving drunk. In the view of the Laramie police, this is a social practice that invites recklessness. For the last few years the Tour has been called off, but not because of the police or a new reverence for sobriety among twenty-one-year-olds. For the last few years the early May weather has been cold, cold and rainy, or cold and snowy.

While durability defines residents of Laramie, even college kids get discouraged biking through the cold slush, so they ditch the bikes and either walk to the bars or go to a party at someone's house. I understand their feeling of defeat at this time of the year. Leaves will never show up on trees; flowers won't poke up from the garden's permafrost: it is hard to believe otherwise. I should remember, but never do, that watching golf or baseball broadcasts from places like Atlanta is masochistic. Only in other places are people wearing shorts, sipping lemonade, mowing lawns.

The high today is 43, with a low of 33. And although when I squint I can see the branches of my tall cottonwood tree taking on a hint of spring suppleness, I don't stop to give it a long look. Instead, I hover over the furnace vent and button my cardigan sweater. May is very much like the March and April that preceded it. I do my best to forget about the weather during the demanding academic semester. With this school year over, though, weather conditions became important to me as I prepared for my trip to the Cooper Nuclear Station, a power plant in Brownville, Nebraska. I'd be leaving behind the college students who, for the most part, make good decisions around risk and heading to a place where a type of energy is created in spite of its representing the highest risk of all commercial energy forms, and which leaves us wondering what to do with the byproducts of its waste.

Wyoming has plenty of coal-fired power plants and wind farms, but no nuclear plants. Very few exist in the arid West, which lacks the quantities of water required to operate one. However, we do have uranium mines, which provide the fuel for nuclear plants. That is, we had them until the 1980s, when the cost of uranium and its dwindling market made commercial mining unprofitable. That left mining towns like Jeffrey City in central Wyoming wondering what to do with the school and houses and stores and other businesses built on the uranium economy. That small town lost more than 90 percent of its population during the bust.

In recent years talk of bringing back that mine and others has surfaced. A system of in-situ mining, or mining the uranium in place, has been developed since those days of the last uranium boom. Heavily involved in statewide energy enterprises, the University of Wyoming surveyed 935 people by telephone in February 2009, 60 percent of whom said they supported the idea of in-situ uranium mining. The strongest supporters were those who described themselves as knowledgeable about how the process worked. In-situ involves dissolving minerals underground and pumping uranium-bearing water to the surface. After the uranium is removed, the water is cleaned and recycled back into the aquifer.

Various environmental and landowner organizations are speaking out with varying degrees of objection about the mining. They voice concerns about water use and whether the mining companies can be trusted to self-regulate. For example, they ask, how "clean" does the water actually get? Bob Gregory, a uranium and industrial metals specialist at the Wyoming State Geological Survey, told me that the process is safe both environmentally and for the workers at the mine and processing plant. "Most of the uranium-bearing fluids are kept within various tanks, vessels, pipes, and so on, and so workers' exposure is kept to a minimum," he said. "They are also checked regularly to monitor their exposure to radiation and have a good track record of safety for their employees. It's also quite safe environmentally, from my understanding," he continued. "Monitor wells are required to ensure that injected water stays where it is intended to stay within and around the ore body."

Gregory also spoke to the manner in which water is used for the in-situ process. "They use water already in the rock formation containing the uranium ore body," Gregory said. "They pump some to the surface, treat it with oxidizers such as oxygen, carbon dioxide, and bicarbonate, typically, and then re-inject the same aquifer water again." Gregory explained that these aquifers are not drinking-water sources "due to the water chemistry and processes that led to the accumulation of the ore body in the first place."

A flurry of permit requests from developers have been made to the Wyoming Department of Environmental Quality (DEQ), and some fear the paperwork load itself will make the DEQ less available for monitoring the actual mining. But with Wyoming sitting on the largest supply of uranium in the country, and the price of that uranium attracting developers, the state is planning to increase the DEQ's workforce. After all, 20 percent of electrical power in the United States comes from nuclear power plants, according to the Energy Information Administration, but the vast majority of uranium is imported from other nations. The Nuclear Energy Institute (NEI) reports that uranium is one of the most abundant minerals on earth, with enough to last at least another one hundred years, at present consumption levels. With more efficient reactors, the supply could last 2,500 years. And while only about 5 percent of the uranium supply is in this country, large supplies exist in nations the NEI describes as stable, such as Australia and Canada.

However, most people's concern with nuclear power doesn't lie at the front end, the mining of uranium. It lies at the back end, in the creation of nuclear power, and the disposal of radioactive waste. That's why I wanted to travel to the Cooper Nuclear Station power plant in Nebraska, to see the process for myself.

Brownville is a town of fewer than two hundred people on the banks of the Missouri River. I plotted almost the entire trip on back roads, including U.S. 30 from Wyoming across most of Nebraska. Trucks along the parallel route of I-80 zipped past me doing at least seventy-five miles per hour, while I happily putted along doing sixty. I'm saving fuel, I told myself, watching the passing of time barely register on the dashboard clock.

My contacts at the plant suggested I stay in Auburn, ten miles west of Brownville, where I could choose among several motels. Just across the river in Missouri is the town of Rock Port, population 1,395. On a ridge there four wind turbines generate enough electricity to power the town. That is, the turbines generate the equivalent of what the town uses, but the town does not literally draw all of its energy from those wind turbines. Wind power is too intermittent, so in reality they receive power generated from many sources we all use, such as from nuclear plants like the one across the river.

If one were to perch on top of one of those tall wind towers, one could see west into Nebraska across the brown, treacherously channeled Mighty Mo. The town of Brownville would be visible. The Cooper Nuclear Station power plant there is capable of powering far more than just a small town. Cooper's net generating capacity is just about 750 megawatts. Geographically juxtaposed as these two types of energy production are, they might serve as a handy graphic for our nation's energy portfolio. But as the national debate rages between sprawling but zero-emission wind farms and smaller but carbon-dioxide-belching coal-fired power plants, nuclear energy has raised its hand and asked to speak. And it is saying that nuclear energy is green, too. What they mean is that the plants emit no greenhouse gases because they don't burn anything to create electricity. That sits well with British scientist James Lovelock. He's the one who developed the theory of Gaia, which says the earth is a living organism. Lovelock argues that global warming, caused in great measure by human activity such as the burning of fossil fuels, is a calamity greater than any civilization has faced so far.

According to Lovelock,

> What makes global warming so serious and so urgent is that the great Earth system, Gaia, is trapped in a vicious circle of positive feedback. Extra heat from any source, whether from greenhouse gases, the disappearance of Arctic ice, or the Amazon forest, is amplified, and its effects are more than additive. It is almost as if we had lit a fire to keep warm, and failed to notice, as we piled on fuel, that the fire was out of control and the furniture had ignited. When that happens, little time is left to put out the fire before it consumes the house. Global warming, like a fire, is accelerating and almost no time is left to act.

Lovelock doesn't dismiss renewable power sources but argues we are too close to the breaking point of climate-change disaster for them to do any real good.

> If we had fifty years or more we might make these our main sources. But we do not have fifty years; the Earth is already so disabled by the insidious poison of greenhouse gases that even if we stop all fossil-fuel burning immediately, the consequences of what we have already done will last for a thousand years. . . . By all means, let us use the small input from renewables sensibly, but only one immediately available source does not cause global warming and that is nuclear energy.

Lovelock's reasons for embracing nuclear power are not those touted by most of its advocates. But even those who do not share Lovelock's doomsday scenario acknowledge nuclear power has other advantages. It is a fact that nuclear plants take up a small fraction of the space needed for fossil-fuel-burning power plants, and a smaller fraction of the space needed for a wind farm producing comparable amounts of electricity. But after an accident at Pennsylvania's Three Mile Island nuclear power plant in 1979, when a partial core meltdown led to the release of radioactive gases, nuclear energy lost its appeal for many Americans.

Nevertheless, there are 104 nuclear reactors operating at sixty-five sites without much fanfare in the United States. There are four different reactor vendors, twenty-six operating companies, and eighty different designs. Nuclear power produces about 20 percent of the electricity created in the country today, according to the EIA. In 2010, U.S. nuclear plants generated 807 billion kilowatt-hours, which is an increase of 1 percent in annual nuclear generation, up from 806.2 billion kilowatt-hours in 2009. Illinois ranks number one in nuclear power generation, followed by Pennsylvania, South Carolina, New York, and Texas, according to the EIA. Nebraska ranks twenty-third.

Before my visit I wanted to be sure I understood in detail the objections to nuclear energy as a power source. I talked to Andrea Shipley, executive director of the Snake River Alliance based in Boise, Idaho. That organization is a nuclear power watchdog group and vocal clean-energy advocate. Shipley left no ambiguity in conveying her opinion about nuclear energy. "Nuke power uses too much water, costs too much, leaves hundreds of thousands of years of radioactive waste, poses serious safety concerns, and will never be built fast enough to solve any of our problems."

Shipley encouraged me to ask questions during my visit to find out what happens to the spent nuclear fuel. She reminded me that security of nuclear plants is a major issue that cannot be undertaken lightly. She suggested I ask about groundwater contamination, and other health issues. So, armed with healthy skepticism, a long list of questions of my own, and a desire to better understand the process and culture of nuclear energy, I was ready for my tour.

I was scheduled to take a tour led by Mark Becker, media relations specialist with the Nebraska Public Power District (NPPD), which owns the plant, and Glenn Troester, communications coordinator at the plant. As public relations representatives, their e-mail correspondence with me was cordial and welcoming. I know it is their job to make a good impression on inquiring journalists and they were very responsive to my tour request—in fact, quicker than anyone else I visited who represent less controversial forms of power.

NPPD provides not just nuclear energy, but also electricity from wind, from coal- and gas-fired plants, and even hydroelectric dams for its customers. The Cooper plant is managed for NPPD by a company called Entergy. Entergy is a large integrated power company dealing in production, distribution, and marketing. It operates eight nuclear power plants in the United States. As NPPD's media relations director, Becker has a full-time job communicating with the public and the media whenever power is in the news in his state, from downed transmission lines after a storm to rate increases for its customers.

Becker is a journalist by training who has spent much of his career working in media relations for various chemical concerns. His manner suggests he's not made nervous by things that could go "boom." Over burgers and fish 'n' chips at the Wheeler Inn in Auburn the night before my tour, he gave me an overview about nuclear power in general and Cooper, specifically.

Cooper is a baseload plant—it runs at full capacity all the time, unlike the wind farm across the river, for example, which adds power to the grid as needed. Cooper doesn't have the type of domed silhouette I thought of when imagining a nuclear plant. Rather than using pressurized water, like some plants which have that appearance, it uses boiling water to make steam to turn turbines to create electricity. Construction on the Cooper plant began in 1968, and commercial operation began in 1974. Its operating license will expire in 2014. The plant is in the process of renewing its license with the Nuclear Regulatory Commission (NRC). The application includes 1,176 pages of documentation about safety and 649 pages of environmental review. If the application is approved, Cooper could operate for an additional twenty years.

I took a long walk after dinner that evening, past the town swimming pool and walking trail and through some of the chummy nineteenth-century neighborhoods. I strolled along a concrete path that circled Rotary Lake and watched families feed the ducks. Downtown Auburn was quiet on this Monday night so I headed back to the motel to review a packet of material Becker had given me about the Nebraska Public Power District and its nuclear power production.

That night I had a dream in which Jane Fonda and Homer Simpson appeared as characters in her nuclear-accident disaster film *The China Syndrome*. Apparently those two pop culture icons represented in my subconscious how much, or how little, I knew about nuclear power. I awoke early, swigged down some bitter motel-room coffee, and joined the flow of morning traffic on the short trip to the plant. I drove past the small state park that commemorates a Lewis and Clark campsite on the west side of the Missouri River. I arrived at Cooper right on time, at 0800 hours. I had submitted my Social Security number to Glenn Troester a few weeks earlier, so that I could be cleared by the NRC to enter a nuclear power plant. Years ago, before the September 11 terrorist attacks, group tours of nuclear power plants were routine. Even groups of school children were encouraged to visit. But now no one can enter without a photo ID, at a minimum, Troester told me.

I carefully followed the procedures I'd been given before my visit: no synthetic fibers (attracts radon particles); no shoes with metal shanks (sets off metal detector); no lip balm (on lips okay, in pocket not okay); don't forget photo ID; don't bring "knives, ammunition, firearms, fireworks, or anything else that goes *pop*, *zip*, *buzz*, *hummm*, *whirr*, *zing*, *hiss* or even *pffft*." That included my camera and digital voice recorder, two of my most essential reporting tools. I had to rely on my hastily scratched notes, my memory, and Troester's willingness to photograph whatever I asked him to and e-mail it to me later.

Becker, Troester, and I met in the Learning Center, where Troester propped up a cardboard poster as a visual aid to explain how nuclear energy works. Troester looks in part like the former newspaper publisher and editor he once was, and the science teacher he's effectively become. He handed me an information card with a plastic bubble window in its center. Inside the window was a half-inch-long simulated fuel pellet. The real stuff would be a whole lot heavier. Using his best Socratic method ("How does fission work, do you know?") Troester explained the life cycle of uranium. After being mined in places such as Wyoming, it is enriched for fuel using one of a few different processes. It is fabricated and encased in zirconium tubes, or fuel rods. Around thirty-three thousand rods are assembled into 548 fuel bundles, which are contained in the reactor. That's where the U-235 isotope, which fissions readily, joins with the U-238 isotope, basically the slow driver in this process. Inside the reactor these materials bounce and jostle against one another, creating a fission chain reaction. Once the reaction is self-sustaining, it achieves what is known as critical mass. The reaction is tempered and regulated mostly by water.

At the time of my visit, a bundle of uranium cost $645,000. If that pellet in the plastic card had been real uranium, the energy it could create would be equal to 149 gallons of oil, a ton of coal, or seventeen thousand cubic feet of natural gas. I thought about the tons of Powder River coal that leave Wyoming every day to be burned in fossil-fuel power plants. Much of Wyoming's Powder River Basin could still be part of the ground near Gillette if uranium, rather than coal, was the fuel of choice.

Radioactive fuel creates waste. One might see tractor trailers marked with the radioactive symbol on the interstate, hauling waste to a disposal site. According to the NRC, most of the waste those trucks carry is low-level radioactive waste (LLRW), such as contaminated shoe coverings, mop heads, and so forth used around hospital radiation units. The items contaminated with LLRW are stored for a time, typically on site, then disposed of as ordinary trash or shipped in approved disposal containers to one of many sites around the country. This system is subject to approval by the U.S. Department of Transportation and the NRC.

High-level radioactive waste (HLRW), such as spent fuel from nuclear plants, is another matter. Trucks aren't hauling it across the interstate system because there is no permanent place for it to go. In 2002, commercial nuclear plants were storing an estimated forty-six thousand metric tons on site. By 2008, that number jumped to an estimated fifty-seven thousand metric tons. The NRC expects an increase of two thousand metric tons each year of stored waste. When people take out their trash at home, usually someone else totes it away to the landfill. And unless that someone else falls down on the job, the matter of smelly garbage at home is efficiently resolved. But in the case of HLRW, suddenly there is no landfill to take it to. One can only bury so much garbage in one's backyard. The plan to create a central depository for used fuel at Yucca Mountain, Nevada, has been aborted by the Obama administration, although whether the end is really the end is yet to be known. The conversation continues.

Meanwhile, each plant is up to its own devices for figuring out where to store used fuel. On my tour of the Cooper plant, we took up that issue on our first stop. Peering out the large windows of the plant's Learning Center, where our tour began, Becker, Troester, and I could see the used fuel dry cask storage area, Cooper's "temporary" solution to the problem. This man-made Yucca Mountain of concrete and steel looked anything but temporary, connected not just to the ground but to the bedrock of the Missouri River below.

The dry casks are typically steel cylinders that are either welded or bolted closed. The steel cylinder provides leak-tight containment of the spent fuel. Each cylinder is surrounded by additional steel, concrete, or other material to provide radiation shielding to workers and members of the public. The used fuel in this dry cask must first be cooled for at least a year in the used fuel pool. That's basically a thirty-eight-foot-deep swimming pool for the extremely hot and highly radioactive fuel rods that have been removed at the end of their lifespan.

Before we could see the pool or other places in the radioactive control area (RCA), Becker, Troester and I had to pass through security. As one might hope, security at a nuclear plant is anything but casual. It starts with the approach to the plant and the view of the imposing gates topped with intimidating razor wire. It includes the watchtowers in which guards armed with military-grade assault rifles perch. They are authorized to shoot anyone whose actions threaten the plant or the people in it. Additional armed guards are on the ground outside and around the control room at the heart of the plant.

To enter the RCA, we all signed consent forms saying we understood we were entering a place with potential radiation risks, then we went through an airport-style metal detector with our personal belongings slid through an x-ray machine. Then we headed for what amounted to a reception area at the RCA to pick up our direct reading dosimeter, which monitors radiation levels along the tour route, and a "trip ticket" card. Visitors are required to know the normal radiation levels along the tour route, and having us fill out the trip ticket ensures that we know this information before starting the tour.

Signage reminded us of all the advantages of ALARA—"as low as reasonably practicable"—when it comes to radioactive exposure. Troester's goal for his tour charges was an expected gamma ray dose of not more than 0.3 millirem (mrem) for the entire tour. Reassuringly, our expected contamination level was 0.0 (zero). Although Troester had been in the RCA many times during his nine years at Cooper, he estimated that his total accumulated dose from the plant was less than 40 mrem. The NRC estimates that radiation exposure from natural sources in the United States averages about 310 mrem. Natural sources include background radiation from cosmic and terrestrial sources. In Wyoming, I'm exposed to both of these sources at a higher level than average Americans because of the thin atmosphere at our altitude, and because of the presence of radioactive material in the ground, such as uranium. The NRC says that another 310 mrems of exposure the typical American receives is from commercial sources such as medical procedures, primarily CT scans.

The dosimeters the three of us wore were programmed to beep loudly if we entered an area where we would be exposed to 10 mrem in an hour. If a visitor milled around in the RCA long enough to be exposed to 50 mrem of radiation, the alarm would also "holler," Troester explained. Troester said that if someone on a typical tour received even 2 mrem, as tour leader he would have a lot of explaining to do.

So in we went to the RCA, swiping plastic cards like hotel room keys in and out of door locks, being "pulled" into sensitive areas by Troester, and "pushed" out of them when it was time to leave. That means he's the first of us to enter an area, and the last to leave it. We passed down long halls with exposed pipes—blue for water circulation, red for fire protection. We moved toward the reactor building until we reached two heavy steel doors separated by a small room called an airlock. The reactor building is kept at a slight vacuum, and the airlock ensures that air from inside the reactor building does not escape into the hallway outside. We had to make sure one door was completely closed before we could open the other.

When we exited the airlock and entered the reactor building, the first thing we encountered was a magenta-and-yellow "high radiation area" sign on a massive door. This was the gateway to the drywell that houses the nuclear reactor. There were magenta-and-yellow ropes in several areas, with "contaminated area" signs on them. I wasn't sure how I could be safe on one side of the rope but in danger only two inches on the other side of it, but I wasn't going to stand there pondering.

Troester showed us the hydraulic control units that drive control rods into the reactor to shut it down. The units are pressurized, and can automatically shut down the reactor in seconds without any human intervention, electricity, or other external power. Troester explained that this is one of many defense-in-depth redundant systems designed to keep the plant safe. It is also part of what makes a nuclear plant so expensive to build. Even with the extra costs, the NRC's information digest shows that in 2008, production expenses averaged $21.16 each megawatt-hour for nuclear power, compared to $35.67 each megawatt-hour for fossil-fuel plants. Their data show that while operation and maintenance expenses at a nuclear plant are much higher than at a coal plant, the cost of the coal was more than five times higher than the price of fuel at the average nuclear plant.

Building a new nuclear plant takes a major investment of time, regulatory steps, and dollars. According to the EIA, an average 2,200-megawatt dual-unit nuclear plant costs $5,272 per kilowatt-hour. That's compared to $2,409 per kilowatt-hour for a 100-megawatt wind farm, or $2,809 per kilowatt-hour for a 1,300-megawatt coal plant. Costs for nuclear plants also include expenses like waste disposal. In spite of expenses, the NRC says that it expects to review twenty-one combined commercial and operating license applications for approximately thirty new reactors over the next few years. A typical plant takes seven to ten years to build.

Probably it was just psychological, but when my tour group found itself standing right above the dry well, which contains the reactor, I felt like the fillings in my teeth were coming to life. True, there was at least thirty feet—eighteen of it concrete and steel—between the bottoms of our boots and the top of the dry well, with the reactor itself well below that, but it still was an eerie feeling. Troester showed us the used fuel pool, in which almost all the fuel bundles the station has consumed since 1974 have been kept innocuously cool at one time or another. Eventually, those bundles will enter the longer-term, temporary dry cask storage area we'd viewed from the window of the Learning Center.

We made a few more stops in the radioactive area of the plant, with me feeling guilty each time I asked Troester to photograph something for me. He occasionally stopped to read our electronic alarms for radiation dosage levels to make sure we didn't break the creed of ALARA. He pointed out the standby liquid control area, which would coat the reactor with boron if the reactor needed to be shut down and for any reason the hydraulic control units couldn't insert the control rods into the reactor. This is another part of the so-called redundant safety system.

This and all other nuclear plants of Cooper's vintage undergo "aging management" in which pipes, wires, and other mechanical systems must be repaired or replaced before they experience age-related failures. Although Troester and Becker joked that "aging management" could refer to themselves and other members of the Cooper team of a certain age, it is one factor, along with redundant safety, that they argue makes American nuclear power plants some of the safest industrial facilities in the world. It also adds substantially to the expense of operating a nuclear power plant.

Next we moved to the control room, outside of which stood another rifle-bearing security guard. We were not allowed to enter the control room itself, so we stayed in the corridor looking through the large window while Troester described the functions of various panels of knobs, levers, dials, and computer screens. One of the panels was a representation of the reactor, lit up with red lights. Troester explained that when a control rod is withdrawn from the reactor so the nuclear reaction can occur, the light turns red. When the control rods are inserted, stopping the reaction for routine reasons, all the lights are green. So red means on, and green means off. Troester said the control-room panels are set up this way because red signifies that people need to pay special attention. When the reactor is operating, he said, the operators—who are licensed by the federal government—need to be highly vigilant at all times. The red lights on the control room panels are a reminder of this need for vigilance and caution. When the reactor is safely shut down, the green lights mean the plant is in "safe" condition.

Even when it's shut down, the system still produces heat. In order to control heat, Cooper uses feed water. They have plenty of it from the Missouri River right outside the door. Proximity to the river, and the background of technicians working at the Cooper plant after working on nuclear submarines in the navy, leads to their method of naming locations within the plant. Instead of using floor numbers, they describe locations in terms of feet above sea level. The "ground floor" of the

Cooper Station is 903 feet above sea level. The used fuel pool is at 1,001 feet above sea level. We made one of our last stops on the tour at 932 feet, at the turbine operation deck, which is really the point of the whole place—that's where the electricity is created. Cooper employs two low-pressure turbines and one high-pressure turbine, driven by steam from the reactor. The steam carries radioactive materials with it, so the turbine generator system at a boiling-water nuclear plant is inside the RCA.

Because of the radioactive gases, a one-hundred-meter stack outside at the Cooper plant releases the gaseous form of plant emissions into the atmosphere. Some of these gases are inert; others are radioactive, although they may also occur naturally, like radon. Troester explained that the release of radioactive gases is strictly limited by federal and state permits and is closely and continuously monitored. "Don't let anyone tell you that nuclear power plants don't release radioactive gases," he said. "They all do it."

After about forty-five minutes in the radiation area of the station, it was time to head to a safer area of the plant. We repeated the complicated sequence of door openings and closings and electronic pass-card readings. Then we reentered the RCA access point and stepped into tube-shaped personnel contamination monitors that carefully scanned our clothes and our exposed skin for even the faintest trace of contamination. Even common radon gas would cause the PCM to refuse to let us leave the RCA, Troester told us.

After we gained the approval of the PCM machines, we turned in our dosimeters and learned that our radioactive dose for the tour was 0.035 mrem, just a fraction higher than Troester's 0.030 mrem goal for this tour, but still below the 0.06 mrem of a chest x-ray, or even the 0.01 mrem of one bitewing dental x-ray, Troester explained.

Back outdoors in the high-security protected area, we saw a maze of fences designed to slow down anyone seeking to enter the station by force, enough so that security officers could detect their presence and open fire to stop them. Also outside we saw huge blue tanks that hold crystal-clear water used to replace water lost in the boiling process. Troester explained that an extensive array of high-tech resins constantly filter and polish the reactor system water to remove radiation and impurities. Water from the Missouri River is used to condense steam back into water, and to keep the many safety systems cool. River water and the pristine-pure reactor-system water are not allowed to mingle, and each water type is in its own isolated and self-contained system of pipes, valves, and pumps. After the river water absorbs heat from the plant, it is filtered with the goal of removing any trace of radiation or contamination, and then discharged back into the river. The discharged water is supposed to meet strict federal and state temperature limits to protect fish and other aquatic life in the river.

My tour team stepped inside the intake building (where yet another heavily armed security officer was posted) to see how cooling water from the Missouri is drawn into the plant through a screen system designed to keep out fish, logs, and various detritus. A fish diverter enables live fish and turtles to escape. But anything unable to swim winds up in a large, very smelly bin, which on this day was littered with several fish carcasses.

We made our way back to the secure reception area, where we turned in our green visitor hard hats, safety glasses, visitor badges, and various gizmos, and talked about what life would be like at the plant in the future. Troester will be busy with the extra visitors from the NRC because of the license renewal process. And he's preparing for the next refueling outage. During the outage, approximately 700 additional workers will flood the area, joining the 750 already working there. The primary purpose is to rotate out used fuel and replace it with new. During the outage other inspections and maintenance tasks will be performed. For about thirty-five days, NPPD will have to replace the power from Cooper with other electricity generated in Nebraska or neighboring states. Outages are planned during times of lower demand from air conditioning and agricultural irrigation. This will likely be Troester's last outage before retirement. He won't have to deal with plant tourists during this period, though. "Management doesn't like to have anyone extra hanging around here during that time," he explained.

Bob Engles, who has been Auburn's mayor for about seven years, has experienced several of these planned refueling outages. He recalls that before Entergy took over management, outages could last sixty or ninety days. He said he took that as a sign there were some serious problems in the plant needing repair. But now the outages last closer to thirty days, which he interprets as meaning that "the people running things know what they are doing."

Engles said that when so many workers come to town for the outage, residents scramble to find room for them to stay. The motels and RV parks fill, and many people rent out rooms in their homes. The town offers its city facilities to plant management for a base of operations for temporary employees to complete paperwork, undergo drug testing, and other preemployment necessities.

An Auburn native, Engles is a lifelong resident of southeast Nebraska who returned with his wife to Auburn for good more than thirty years ago. He owns an insurance and real estate business and says he has a personal opinion about the plant outside of his public role as town mayor. "We are very grateful to have the plant here, not just for the economic value but for the commitment those people have to our area. They are great partners," he said.

Engles suspects that hosting a nuclear plant brings his town and surrounding Nemaha County certain elevated security risks. But he believes most locals aren't terribly concerned with plant operational safety or whether the town is vulnerable to terrorist attack. He thinks they feel comfortable about the operation of the plant because many of them are employed there and see how it operates firsthand.

In August 2010 the NRC issued its final supplemental environmental impact statement for the proposed renewal of Cooper's operating license. The report concluded there were no environmental impacts that would preclude license renewal for an additional twenty years of operation. This isn't the same as allowing the renewal, but it is a significant step in a long process. If the license renewal is accepted, the plant can operate until 2034. Still, that means the town of Auburn will someday have to face a post-Cooper economy. Engles can imagine Auburn continuing to host a power plant, possibly even a nuclear plant. "The need for energy in this country isn't going to be any less then than it is now," he said.

Mark Becker, with NPPD, said public comments taken so far during the license renewal process have been supportive. Though the plant has been written up for what Becker describes as minor safety violations, these violations occurred before management of the plant was taken over by Entergy. However, the NRC has observed several noncritical safety violations at the plant since that time. In 2008, the plant received a more serious "white significance determination," prompting closer ongoing inspections from the NRC. Victor Dricks, with NRC's Region 4, where the plant is located, told me that "the plant has taken care of the problem" that caused that level of scrutiny. Further, he said the problem would not impact the plant's application for license renewal. "That is based on the plant's ability to manage aging," he said.

In June 2011, significant flooding along the Missouri River prompted the Army Corps of Engineers to release water from two dams upstream of two NPPD nuclear plants. The Fort Calhoun nuclear plant near Omaha happened to be closed for refueling during this event, but officials from the NRC, the Army Corps of Engineers, NPPD, and other organizations were on hand to monitor events. The Cooper plant was operational during the flood, and although the Missouri's waters were high, plant operations were not in danger, according to plant officials. However, whenever flooding occurs in the area, the plant issues a Notification of Unusual Event, and this time was no exception. This is the "lowest and least serious of four emergency classifications established by the Nuclear Regulatory Commission for nuclear power plants," according to the NRC. At this classification level, additional internal doors are closed. At the next level, the plant would be shut down, as a precaution.

In 2011, the NRC performed additional checks on all U.S. nuclear power plants in response to the earthquake and tsunami disaster affecting the Fukushima Daiichi power plant in Japan. As a result, Cooper Nuclear Station will receive additional oversight because NRC inspectors reported that the plant's procedures to manually operate valves that are part of the high-pressure coolant injection system would not work properly in the event of a fire. The findings at Cooper during these nationwide inspections were reported at other plants, too. The NRC sums up its findings about plants nationwide:

> While individually, none of these observations posed a significant safety issue, they indicate a potential industry trend of failure to maintain equipment and strategies required to mitigate some design and beyond design basis events. Nuclear plants have multiple, redundant, strategies for which the overall function is to mitigate damage to the facility's fuel elements and containment. The failure of a strategy due to equipment failure, procedure inadequacy, inadequate training, etc., does not mean that the other redundant strategies would not have successfully performed their function. During this inspection, while some deficiencies were identified that would have caused a single strategy to be compromised or fail, no functions were compromised that would have resulted in damage to the fuel elements or containment.

The NRC is an independent agency created by Congress. Its

mission is to "license and regulate the nation's civilian use of byproduct, source, and special nuclear material in order to protect public health and safety, promote the common defense and security, and protect the environment." The final item on this list has been the magnet for public scrutiny in recent years, especially by environmental groups such as the Idaho-based Snake River Alliance. Even with the NRC positioned as the referee there is still a lot of sparring going on. Perhaps as a result, there is a beleaguered air around many in the nuclear power industry.

One industry observer who didn't want to be named told me the power companies and regulatory agencies are used to being described as "lying bastards." On the other side, antinuclear activists are dismissed as folks who "never met a neutron they liked." Even Mark Becker takes a bit of a swipe at the environmental movement. He compares the nuclear industry's record to actions taken by Ira Einhorn, an early environmental activist involved in the founding of Earth Day. Einhorn is serving a life sentence for the murder of his ex-girlfriend.

"Ira Einhorn has killed more people than a nuclear plant ever has," Becker pointed out.

Andrea Shipley, for her part, swipes back. Her Snake River Alliance argues that no nuclear power is good nuclear power. She starts with the question of license renewal.

"For a plant to stay open another twenty years past the original decommissioning date is a cause for concern around safety," Shipley said. Eliot Brenner, director of the NRC's Office of Public Affairs in Washington DC, said the NRC understands the concern about aging but gives it little credence. "About the only thing forty years old when a plant's license is renewed is the ink on the original license," he said. "Plants routinely replace safety-related systems and major components as part of refurbishments as components age, and technology improvements are often built into the refurbishment plans to keep technology current. We have one mission—protecting people and the environment—and that is at the core of every decision we make for plants in general and license extensions in particular," Brenner said.

In addition to concerns about aging, Shipley has doubts about the safety culture in nuclear power plants. She offers evidence of a 2002 survey of the NRC's workforce, commissioned by the NRC's Office of the Inspector General (OIG) and conducted by an independent contractor. Shipley said it revealed "troubling facts about employees' confidence in the agency's ability to be an effective regulator." Many employees reported a concern that the "NRC is becoming influenced by private industry and its power to regulate is diminishing." Meanwhile, Shipley said, only slightly more than half of NRC employees reported feeling that it is "safe to speak up in the NRC"—a finding that does not instill confidence in the NRC's ability to identify potential safety problems before they become serious.

But Brenner asserts right back on this issue. "It is a bit of a stretch to attempt to link a seven-year-old study on the attitudes of NRC employees with the safety of plants whose licenses have been extended." He pointed to safety-culture studies that appear to show increased confidence by NRC employees "in the agency's ability to regulate uses of nuclear materials." He said the NRC has been rated two cycles in a row as the best place to work in government, "in part because of the way its managers listen to employee concerns." His interpretation of this information is that it is "the hallmark of an agency that knows what it is doing and has confidence in its abilities."

As I considered that debate I thought about Shipley's suggestions that I inquire about groundwater contamination. According to Shipley, "the nuclear industry has recently come under fire for leaking tritium—a radioactive isotope of hydrogen—into the groundwater of areas surrounding nuclear plants. Even worse, nuclear energy companies have kept the discoveries of these leaks from the public, sometimes for several years."

Naturally, Brenner sees things a little differently.

“The EPA allows up to 20,000 picocuries of tritium per liter of drinking water,” he said. “In most cases the tritium found at plants has been on plant property and plants have moved rapidly to institute remediation efforts where tritium has been found. Tritium, if ingested, can be rapidly flushed from the body by simple hydration and subsequent urine flow. The NRC pays close attention to the issue of leaks of tritium or other substances into groundwater at plants.”

The EPA’s explanation of tritium acknowledges its dangers but does not accuse nuclear power plants of covering up when a leak occurs.

> In the mid-1950s and early 1960s, tritium was widely dispersed during the above-ground testing of nuclear weapons. The quantity of tritium in the atmosphere from weapons testing peaked in 1963 and has been decreasing ever since.
>
> Today, sources of tritium include commercial nuclear reactors and research reactors, and government weapons production plants. Tritium may be released as steam from these facilities or may leak into the underlying soil and groundwater. However, such releases are usually small and are required not to exceed federal environmental limits.

Because tritium glows, it is used in signage over building exits, for example. According to the EPA, when someone improperly throws such a sign into a landfill and it breaks, “water, which seeps through the landfill, is contaminated with tritium from broken signs and can pass into water ways, carrying the tritium with it.”

Shipley does not specifically indict the Cooper plant with accusations of a hostile safety culture, sloppy safety, or a poor environmental record. She is equally concerned about all nuclear plants because after all, if something went really wrong, people could die. “While the staff may be capable and trained, there is no room for mistakes within the nuclear industry,” she said.

Brenner agrees that there is no room for mistakes and said, "The safety and reliability of plants has been on the increase since the 1970s when plants were running at capacity levels in the 60 percent range. Today, thanks to rigorous NRC oversight and an industry consolidation that has led to far better, standardized management, plants are running in the vicinity of 92 percent of the time. All safety trend indicators continue to show declining incidences of safety-related actions or incidents at plants, again due to NRC oversight and sound management."

So which position is correct? Arguments one hears in favor of nuclear emphasize what it doesn't do. It doesn't take up as much room as other forms. It doesn't create greenhouse gases. It has avoided most of the safety hazards once feared. In Gallup's annual Environment Poll, a steady number of Americans support nuclear energy. Since 1994, when Gallup first began asking about nuclear energy, support for its use has been fairly steady in the mid–50 percent range. In 2011, just after the accident at the Japanese nuclear plant, 58 percent of survey respondents thought U.S. nuclear plants were safe, and 36 percent thought they were not. Gallup suggests that although this survey took place shortly after the Japanese disaster and respondents might not yet have processed how they felt about the events, support for nuclear power in the United States has held steady for the past ten years. In spite of that, the majority of Americans are reluctant to build a power plant in their vicinity. There are twenty-nine commercial nuclear power reactors in the United States that were formerly licensed to operate and are now permanently shut down. Another sixty-four planned nuclear power projects have been canceled. No new nuclear plants have been built since Watts Bar 1 in Tennessee, in 1996. That leaves the existing plants, like Cooper, to get its license renewed or close, taking the power it contributes to the grid with it.

After taking my leave of Becker and Troester at the Cooper Station, I drove the short distance to the state park at the edge of Brownville. I got out of the car with my Nebraska map and sat on the west bank of the Missouri River, in the shade of the Meriwether Lewis River Boat. It sat beached and delinquent near a historical marker explaining that Lewis and Clark's Corps of Discovery had camped here on their way up the Missouri. They'd traveled about ten miles that day, Sunday, July 15, 1804. Clark's journal describes his swim around various side creeks waiting for the boats to catch up, and his explorations of the area when he did so. He describes seeing timber, hills, plains, and a few deer. He saw "Great quantities of Grapes, Plums of 2 kinds, Wild Cherries of 2 Kinds, Hazelnuts and Gooseberries."

Remarkably enough, the area seems much the same today, if you ignore the power plant, the wind farm, and the bridge spanning the river between the two. Bob Engles, the Auburn mayor, cautioned that "unless they want to drown," nobody swims in the Missouri anymore. "It is much more treacherous than it was two hundred years ago," he said. Waxing philosophical, he added, "Do you realize the Lewis and Clark explorers maybe did not even understand the concept of electricity and lights? Now we have nuclear fission splitting atoms at a nuclear power plant at almost the same site they swam. What do you suppose, in their wildest dreams, they thought would develop in this country? What will be at that site two hundred years from now?"

4

Solids, Liquids, and Gases

A Texas Gas Field

JUNE 1

For each of my birthdays that I can recall, which is about forty-five of the present fifty, cold wet rain has been the order of the day. Anyone who doubts me might think back to what they were doing on Memorial Day weekend this year. Or the years before that. Were you camping, huddled in a tent hoping you could get out long enough to brew your morning coffee between downpours? Were you stuck in the house, watching another red-flag rain delay at the Indianapolis 500? Welcome to my world.

Technically, my birthday is not on Memorial Day, but it was until the official holiday was moved to the last Monday of May. As one who likes to make the best of things, since the Great Holiday Rearrangement I treat all the days from Memorial Day to May 31 as my birthday season. Depending on when the weekends fall, I sometimes stretch my personal celebration out until the first of June. This year my birthday season was filled with gusty winds and rain. The high on June 1 in Laramie was 62, and the low 44.

I confess to looking forward more than casually to my planned trip to Oklahoma City later in the month. I wanted to go somewhere warm. I was even willing to accept hot, just to get out of the cold. I was planning a tour of a to-be-determined Oklahoma gas field. I'd had several phone conversations with a fellow named Wayne who worked for an oil and gas organization in the state. Although he seemed a bit uncertain who I was and what I wanted each time I called him, he vaguely assured me that we'd be able to load up in a pickup and make a run out to "the patch" to see a drilling rig or at least a producing well any time that was convenient for me.

"Sure, just give me a call and we'll set somethin' up," he said.

Less than a week before my visit I was feeling a bit uneasy about my half-baked plans with Wayne. Reaching him on the phone on the day he'd asked me to confirm, my fears were realized. He was going to be tied up on the day we were to meet and suggested I find someone else. To his credit, Wayne gave me the names and phone numbers of several of his contacts in the Oklahoma oil and gas industry. To my great relief, one of those contacts was someone both helpful and sharp as a tack: Alesha Leemaster, senior communication specialist with Devon Energy. Although she couldn't arrange a visit to a drilling site with such short notice, she did set up a meeting for the two of us and Chip Minty, manager of media relations, at the corporate office in downtown Oklahoma City.

There are people working in media relations today who only know the basics of what their employers actually do. These media "flacks" as they are derisively known among journalists are hired to do one thing: deflect inconvenient information that seems headed for the public. Ideally, and often though, these public relations folks do an excellent service: they are knowledgeable and efficient at getting information to journalists in ways both timely and thorough. In Leemaster and Minty I found the best communications staff I could have hoped for. What I learned from them is all part of the information stew, along with all the other material people have delivered my way. And I knew that without them as my guides, no amount of solo traipsing around a gas field would teach me a single useful thing about how electricity can be the end product of a colorless, odorless gas.

As I prepared for my trip to Oklahoma I realized I had been there before, but not for about twenty-five years, and I had never traveled there by air. It was astounding to see the bird's eye view of the extensive oil and gas fields when I was in the air above the state. Peering out the plane's small window, at first I thought I was seeing individual homes each resting on its own private cul-de-sac. That was until I saw that those short roads leading from main roads were not adorned with houses, but with well pads.

I took a cab from the hotel where I was staying into downtown Oklahoma City, to Devon's headquarters. The tall office building took up most of a block but Devon was not the sole tenant of the building. It has since broken ground on a fifty-plus-story headquarters building in the downtown which, when completed, will be the tallest building in the state.

Devon Energy's significant presence in the city seems suited to a company that is the nation's largest independent natural-gas and oil producer. More than 90 percent of its production is from North America. The company's production mix is about 65 percent natural gas and 35 percent oil and natural-gas liquids, such as propane, butane, and ethane. Devon says it produces over 2.5 billion cubic feet of natural gas each day, but that's still only about 3 percent of all the gas consumed in North America. Devon got a big return on its investment after locating coal-bed methane gas in the San Juan Basin of New Mexico. These days Devon is focused closely on the Barnett Shale "play" in the Fort Worth Basin. This natural-gas resource is referred to in the industry as "unconventional" because the gas comes from shale, not sandstone, which is the conventional source of natural gas.

Living in Wyoming made me familiar with the oil and gas

industry, from exploration and development, to drilling, to production. In the central part of the state drilling rigs are as plentiful as dots on a Dalmatian and they attract the interest of governmental and environmental groups concerned about the effect of drilling on air and water quality and wildlife. But even at its most vigorous, the drilling activity in Wyoming is nothing compared to what it is on the southern plains.

Texas leads the United States in proved reserves of natural gas. According to the Energy Information Administration (EIA), it has nearly a third of the nation's supply and in 2009 produced 6.8 million cubic feet of natural gas. Wyoming was the second highest producer but the drop between the two was steep: only 2.3 million cubic feet was produced there in 2009. Oklahoma is the third top producer of natural gas, though less than a third of its gas stays with in-state customers: most is piped to trading hubs in Kansas and Texas for distribution. Texas consumes more natural gas than any other state. The industrial and power sectors in Texas help the state consume about 20 percent of the natural gas in the United States.

The Baker Hughes Rig Counts is the place to find data concerning active drilling rigs around the world, whether that rig is looking for oil, gas, or even geothermal resources. Baker Hughes has kept records of weekly rotary rig counts since 1944. I randomly chose September 2008 and found that in that month, Wyoming's average number of active oil and gas drilling rigs was 80. By contrast, Oklahoma's was 202. To the south, Texas was averaging 950 active rigs per week, not counting those located off shore.

Drilling numbers started to decline in the spring of 2009, all around the country. For example, Texas's rig count dropped to 380, exactly one year after Baker Hughes counted 950 rigs. The reason wasn't that oil and gas could not be located but that the price, particularly for natural gas, was in freefall. The global economic downtown was one reason for the lower prices. In addition, recent public efforts toward conservation meant that while gas was being developed and pumped and delivered to the market, industrial and residential customers used less of it. A certain amount of natural gas can be stored, but even that supply had exceeded reasonable amounts for the market to absorb.

The resulting glut in the market meant cheaper natural gas, great for those of us who use it to heat our homes or who get electricity from gas-fired power plants. But not great for economies in states like Wyoming, Oklahoma, and Texas, which rely on the revenues from selling natural gas. What a difference a few years makes. In December 2005, the natural-gas city gate price in Wyoming was high, around $9.43 per thousand cubic feet. "City gate" refers to where the company that sells natural gas receives it from the pipeline company. In June 2009 that price was down to $3.32 in Wyoming. By comparison, in December 2005, Oklahoma's price was $11.39 per thousand cubic feet, but in June 2009 it had dropped to $6.68. In December 2005, Texas's price was $10.38 per thousand cubic feet, and in June 2009 it had dropped to $4.25. Although the loss in Oklahoma and Texas was much less than in Wyoming, it is no wonder drilling slowed in those place, too.

The direct and indirect loss of jobs and revenue to the state was significant for many of us in Wyoming, especially coupled with the broader economic downturn. At the same time, some of us breathed a sigh of relief that there was a bit more elbow room on the crowded highways and that the unexpected population booms in surprised small towns had eased back. But the environmental degradation of boom places like Jonah Field and the nearby Pinedale Anticline, both in the Upper Green River Basin, would need much more than a slowdown to repair.

In 2009 the Government Accountability Office (GAO) reported that the Bush administration's 2005 Energy Policy Act, which was intended to streamline oil and gas development, had led to shortcuts around environmental laws and air pollution in several western locations. One of those locations: the town of Pinedale, neighbor to the Jonah and Pinedale Anticline fields in the Upper Green River Basin. The 2005 Energy Policy Act allowed Bureau of Land Management (BLM) officials under certain conditions to sidestep provisions in the 1969 National Environmental Policy Act requiring analyses of threatened or endangered species, historical or cultural resources, human health, and safety of potentially significant cumulative environmental effects of oil and gas drilling on public lands.

An exemption was granted for a particular drilling site and then that exemption was applied to drilling on all BLM lands, in a one-size-fits-all approach. In its report the GAO said these exemptions from normal permitting procedures were "inappropriate." The exemptions were pushed through to speed energy development in the West, although the GAO found no evidence this result was achieved. According to the report, about 25 percent of drilling permits issued from 2006 through 2008 allowed the exemptions. Nearly two-thirds of all the exemptions came from three BLM field offices: Vernal, Utah; Farmington, New Mexico; and Pinedale, Wyoming.

Environmental activists, such as Erik Molvar, with Biodiversity Conservation Alliance (BCA) in Laramie, have long directed the fight not so much against gas drilling but against well density. In 2009, the group sued the BLM, which grants permits to companies drilling in the Jonah Field, such as EnCana, the largest of those companies. That was in response to a BLM plan, approved two years earlier, to allow more than three thousand wells in the field over the next seventy-five years.

"With its tangled webs of roads, pipelines, and drilling sites, the destruction is already so severe that the Jonah Field has become the nation's poster child for drilling gone wrong," Molvar said. Then he put forward a somewhat rhetorical question. "Since the Jonah Field has already been substantially degraded, and the mineral deposits are so rich, why bother fighting over adding another 3,100 wells in Jonah to reach a density of sixty-four per square mile?"

His answer to the question may be part rhetorical flourish but at its heart is an excellent point. "At issue is whether or not the industry should be allowed to completely destroy lands and habitats in its quest for gas, or whether they should be required to develop the gas with the smallest possible impact." Molvar suggests that technology such as directional drilling should be used. With that method, as many as thirty-two wells can be clustered on a single well pad, and then drilled directionally to drain the surrounding gas reservoirs.

Chip Minty, with Devon Energy, distances the company's practices from what has happened at Jonah Field. He acknowledged that the Jonah and Pinedale Fields are "renowned" for the environmental damage that intensive drilling has caused there. Although Devon has some shale gas leases in the Upper Green River Basin and is exploring in southwest Wyoming, they have faced no specific criticism from BCA. Minty says that on Devon's leases they can drill directionally, also described as horizontally, in areas that are appropriate for that approach and still be environmentally responsible. "We can work with the BLM or historical organizations in Wyoming," he said. "Shoot, we don't have to drill in an area that is sensitive for antelope, deer, or sage grouse. We can start three thousand feet away over a ridge, drill down, and go horizontal and get there just as easily."

But in defense of the vertical drilling process used in the Jonah Field by others, Minty said developers there are targeting very defined pockets of natural gas in conventional sandstone reservoirs. "They can't do what we are doing with shale."

What Devon and other companies are "doing with shale" is a practice that until recently was not often economical. It is called hydraulic fracturing, also known as fracking. The process makes it possible to release gas from extremely dense rock like shale, where organic matter deposited three hundred million years ago has transformed into natural gas. The end result could mean an increase of about 39 percent in our nation's gas reserves, according to the Potential Gas Committee, a group of industry, government, and academic volunteers. And that figure assumes that there will be no more technological advances. With so many people extremely motivated to make shale gas pay, you can lay odds advances will keep coming.

Alesha Leemaster explained that horizontal wells and fracturing have made it economical to produce gas from shale formations. However, horizontal wells are also successful in conventional reservoirs, such as sand and limestone and in nonshale unconventional reservoirs, such as coal-bed natural gas. The horizontal well exposes more reservoir rock than a vertical well does, allowing greater production of natural gas from one well.

She added that fracturing is a well-established technology that has been used extensively for the last fifty or sixty years and was used by a company Devon purchased. "Mitchell Energy began applying fracturing to the shale in vertical wells with some success, and when Devon acquired Mitchell in 2002 we introduced horizontal drilling to the shale. Coupled with fracturing, horizontal drilling unlocked shale plays," she explained.

When Devon Energy acquired Mitchell, Minty and Leemaster said, everyone on Wall Street thought Devon was "crazy." In addition to stubborn nonporous rock, they also had to confront the Ellenberger Aquifer, which lies under part of the leases Devon purchased from Mitchell. If drilling goes too deep and hits water, the water seeps into the shale and kills the well. "Devon spent $3.5 billion on the acquisition and had to figure out how to make it pay," Minty said.

Make it pay they did, and I wanted to see how they accomplished that feat. I had heard so many contrary claims, so to keep them straight I divided them up into two categories. In category A were claims that hydraulic fracturing was safe, that it didn't contaminate water, that it didn't pump dangerous chemicals into the ground. There were completely opposite but believable claims in category B. So when Minty and Leemaster invited me to return to the southern plains to see how gas-shale drilling worked, I jumped at the chance. Hoping to hit a cool patch after an especially warm Texas summer, I arranged my visit for mid-September. This time I flew in to Little Rock, Arkansas, made a swing through several southern states by car to see other varieties of power production, and then drove southwest to Decatur, Texas. Devon's production manager, Fred Cornell, drove over from Bridgeport, Texas, to meet me.

Cornell trained as a civil engineer and is now a registered petroleum engineer. We rendezvoused with Leemaster, who drove down from Oklahoma City, at the Kusin's Convenience Store at the Justin turnoff along Highway 287, between Decatur and Rhome. Leemaster brought me the requisite hard hat, safety glasses, ear plugs, and size-six steel-toed boots borrowed from a woman in her office. We loaded up into Cornell's white GMC Yukon and made for the Barnett Shale.

As Cornell drove and I sat in the back lacing my boots, Leemaster leaned over the bench seat and gave me a primer on the Barnett Shale gas play. The core area covers ten counties from just north of Decatur to south of Fort Worth. At the time of my visit it held 11,600 producing wells, about 3,800 of which are operated by Devon Energy. The U.S. Geological Survey estimates the formation holds more than twenty-six trillion cubic feet of natural gas. That makes its potential yield more than twice that of the Jonah Field in Wyoming. Of course, the geologists back in Oklahoma City first have to determine where to drill. But in the rich gas play of the Barnett, it seems, one could not poke a golf tee into the ground without hitting a lucrative well site. Still, since each hole costs a few million dollars to drill, one would like to have a pretty good sense that it will pay off.

"Drilling is our R&D," explained Cornell, referring to the leap of faith necessary each day for research and development in this business. Gone are the days of watching for oil seeping out of the ground, or even shooting at some food like that accidental wildcatter Jed Clampett and having Texas Tea bubble up from the ground.

Cornell explained that the least visible parts of the drilling process had already occurred at the site we were to visit: leasing, conducting engineering studies, and permitting with various state agencies. This land is not federal land. The BLM is not involved in the leasing process here, like it is in Wyoming and other western states. While Cornell said the leasing process in Texas can be complex, there are not nearly the environmental regulations like there are on publicly owned western lands. But the locations come with their own challenges. For example, some of the drill sites on the Barnett are deep within residential areas. Cornell explained that Devon had held the leases for many years but the sprawling suburbs of the Dallas–Fort Worth metroplex have reached those sites and built on top of them. Although the residents own the property, the company holds the leases to the mineral rights beneath the ground. So a cluster of houses may snug up to a neighborhood pocket park but a few yards away behind a fence might be a producing gas well.

Our tour group's first stop was some miles south of the convenience store where we met. In Wyoming, the main thing places have in common is that they are far apart. So I was expecting the drilling site to be out in the middle of rangeland thirty miles from a town. Instead we drove along a highway busy by Wyoming standards, but fairly sleepy for a road so near the heavy traffic of Fort Worth. We turned off the main road and there it was: a drilling site that had already been bulldozed flat and was ready to go.

We climbed out of the Yukon on this pleasingly cool Texas morning and clambered across the muddy truck-formed ruts toward the office trailer. There were about half a dozen men working, employees of various companies subcontracted by Devon. Kyle Waldrop is the "company man" who works for Devon and oversees the operations. Around thirty, with his wiry build, white smile, and dark hair, he reminded me of the late Davey Allison, still my favorite NASCAR driver. He led our little tour group into the office and introduced us to two subcontractors from the drilling company Helmerich and Payne, or H&P for short. Dago is the tool man, originally from Mississippi, with twenty years in the business. Tony is the safety man, with seven years in the business. He gave us a talk about watching our step on the ladders and on top of the drilling rig. Waldrop, who has ten years in the business, gave us an overview of how they were progressing at the site and showed us a map of where the well would be drilled.

It takes about a week to prepare the average drill site, he told us. Because there was ample space for it at this location, a lined reserve pit was prepared. The material displaced by the drill has to go somewhere, at least temporarily. When space is more limited it goes into a tank—an arrangement known as a closed loop system. Regardless, all this waste must be removed, and the pit reclaimed, by law. And although there is plenty of ambient noise from the highway, a sound barrier is erected around the site to contain the construction and drilling noise. A diesel engine operates around the clock to power the equipment. Diesel engines are not known for being quiet.

After the site was prepared, a moving company delivered and assembled the drilling rig itself. That typically takes just a few days. At this site they use a technically advanced type of rig that the men said was much safer to operate than older-style rigs. To illustrate they all held up their hands to show each still possessed his original ten fingers. Finding three roughnecks with thirty fingers among them is apparently unusual. But this new drilling rig "takes the rough out of roughneck," they told me.

Then it was time for our group, now numbering six, to climb up the steep metal stairs into the driller's cabin. That's when I understood the company's requirement, the only such at any of the sites I'd visit, that I wear steel-toed boots. Between the mud, the ruts, and the steep slippery steps up to the rig, I was glad I had protection not just for my head, eyes, and ears, but for my feet as well. I was also glad I hadn't spent my entire summer driving to power plants, but had also gotten some exercise. I didn't want to appear too klutzy or out of shape as "Davey Allison" and the rest of the crew followed me clomping up the steps to the driller's cabin. That's where we found James at the controls. Though he was surrounded by windows affording him views of not just the Texas plains but of the hole where the drill was throbbing, he did most of his work by observing computer monitors. "It's a lot like playing a video game," he said of his work.

The drilling he was doing wasn't a simple matter of aiming straight down into the ground. James could drill eight thousand feet straight down and then curve his bore hole to a horizontal angle to expose it to a greater amount of shale. He was in the midst of making that horizontal curve when we arrived. The drill bits are made of tungsten carbine steel with a diamond cutter, and cost up to fifty thousand dollars. To ease the drill bit into the ground, the bit is lubricated by a substance called "mud." Mud looks like what you would imagine but also contains bentonite, almost all of which is mined in Wyoming. Mud lubricates and cools the drill bit and also keeps the hole from collapsing until a pipe can be inserted to keep things steady. The mud circulates back out of the hole along with the cuttings.

Our group left the cabin and stomped around together on the floor of the rig, approximately one story off the ground. Waldrop showed us where the mud is shaken, de-gased, and de-sanded before being circulated through the hole again. The junk removed from the mud is dumped in the reserve pit. Waldrop had no love for the mud pumps—"they break down a lot." But as to the rig itself, he and the crew praised the advances in technology that mechanized so much of the work.

Waldrop said each drill rig manufacturer is competitive

about making their equipment better and safer than the other manufacturers. "They've each got to do something a little better than the other guy," Waldrop said.

Drilling-rig safety is worth being competitive about. The Occupational Safety and Health Administration tracks the hazards at drilling sites from site preparation, to drilling and well completion, to servicing a producing well, and finally to plugging an abandoned well. The Department of Labor, Bureau of Labor Statistics (BLS), tracks data on workplace injuries and death. They say the most dangerous occupation is the fishing industry, as anyone who watches the reality crab-fishing TV show *The Deadliest Catch* on the Discovery Channel knows. In fact, that program's producers contacted Devon Energy about filming its reality show *Black Gold* featuring the Devon crews. But the men touring our group agreed they didn't want to have anything to do with it. "They make it look like it's just a bunch of cowboys working out here not paying any attention," Waldrop said of the program, which aired without them. "That isn't at all how it is."

The BLS data underscore Waldrop's assertion about the caution workers in this industry must exercise to stay safe. There were 120 fatal work injuries in the oil and gas extraction industry in 2008. "Transportation events" accounted for forty-nine of these. Fortunately, not all injuries are fatal, but the BLS indicated that in 2007, 4,200 injuries occurred in the industry that required time away from work. More than 32 percent of those injuries resulted from a worker being struck by an object.

In spite of the inherent danger of schlepping around big heavy pieces of equipment in all kinds of weather on uneven ground, Waldrop loves his job. He has risen through the ranks, now well above the lowly "worm"—the term for a brand new worker who is given all the worst jobs. Waldrop said he wouldn't want to do work that didn't keep him outside. "There's my office over there," he said, pointing to the trailer. "As you can see, I'm not in it."

After a bit more visiting at the site, it was time to let our tour guides get back to work. We had two more stops ahead of us and hoped to time our visit to reach a site where workers were engaged in hydraulic fracturing of the porous rock. So we loaded back up in the Yukon and headed to the Rodney Wooley 1H site.

If one could wait around for nature to trigger an earthquake, perhaps natural gas trapped many thousands of feet down in stubborn shale source rock would release on its own and float obligingly to the surface. Instead, oil and gas engineers have taken matters into their own hands. Fracturing "stimulates" wells by pumping a solution of water, sand, and a small amount of other undisclosed substances into the well bore at an intense rate. After a bore hole has been created, as Kyle Waldrop and crew were doing at our first stop, a pipe is inserted into the hole. In fracturing, that pipe is opened at various locations along its length and the solution is pumped in. As a result, the rock cracks and the fluid used to make that crack gets even deeper into the rock. The fracture is kept open by inserting material such as sand into the fractures. Then the gas flows more freely back into the well bore hole.

The fracturing solution is 99.5 percent water and sand, Alesha Leemaster told me. Devon uses about 3.5 million gallons of fresh water for fracturing each well in the Barnett. The water comes from local reservoirs, aquifers, and municipal supplies. This is almost a literal drop in the bucket to what Texas uses for industrial purposes, Leemaster said. As a person trained to turn off the water while brushing my teeth, I found myself mulling questions I'd ask at the water recycling station, planned for the end of our tour.

Water and sand can't do the fracturing job alone, in spite of technical advances. They must be helped by other chemicals. Industry spokespeople state that fracturing shale for gas may require the use of acids to help dissolve minerals, glutaraldehyde to eliminate bacteria from the water, sodium chloride, and a variety of substances also found in materials used in products commonly found around the house. Consumer and regulatory groups around the country where fracturing is done are pressuring the industry to release the complete list of chemicals used so their presence can be monitored in surrounding soil and water. The industry is resisting this pressure.

"Companies don't want to tell their trade secrets," according to Jay Ewing, a completions manager with Devon. His job is to attend to the testing and monitoring of wells after the drilling is complete and production from that well is underway. And while practices and chemical recipes vary by the geology of the location, Ewing said that in the Barnett a biocide to kill bacteria in the water is commonly used. The other chemical used is friction reducer, which Ewing explained is a polymer "just like Palmolive or Joy soap you use at home." The industry is highly regulated by a variety of state and federal agencies, he noted, although the Environmental Protection Agency (EPA) continues to adjust its regulations as drilling technology advances.

Ewing was an engineer with Mitchell Energy who began working for Devon when that company purchased Mitchell. He said he was part of the group in the early 1980s that argued that it wasn't economical to recover gas from the shale. He said George Mitchell "wouldn't take no for an answer" so they kept at it. In the early days of the Barnett the conventional thinking in engineering was that a fracturing job needed fluid with fairly high viscosity so that sand would stay suspended in fluid. "Sand in fresh water just falls to the bottom," he said. In those early days fracturing at the Barnett was all done vertically and used "massive hydraulic fracks." This method used chemicals to create viscosity and still other chemicals that were time released to turn that viscous fluid back into water. The chemicals, such as guar gum, were then pumped back out, leaving behind the water.

"Those jobs in the 1990s were more expensive. A single job cost a quarter million dollars," Ewing said. So they tried just straight freshwater and sand, without viscosity-adding chemicals, and discovered it would work. "It makes better wells and costs only fifty thousand dollars. That was the first step in making these shale plays viable. Then Devon bought Mitchell and took it horizontal. Horizontal had been around, but fracture stimulation of horizontal was less common."

"Frack jobs and drilling wells are expensive, especially with shale formations," he explained. Ewing said that the Barnett Shale formation is five to six thousand feet below any rock with drinkable groundwater. The Barnett formation sits above the Ellenberger, which is a large saltwater aquifer, not potable water.

"You want to be certain that the job you are pumping is only going into that formation. That's why geologists look at formations above and below your targeted formation to look at stress within the rock you want to crack and stress of rock above and below. With the Barnett, the stress within rock that has hydrocarbon in it is less than the rock above and below, so it is easier to fracture that rock with the hydrocarbons in it. You have control."

Fred Cornell, Alesha Leemaster, and I were about to see this complex and highly regulated process at work at Rodney Wooley 1H. We pulled in to the work site and were greeted by folks who naturally wondered who we were and what we were doing there. If the drilling site felt like a small village, this fracturing site felt like a mall parking lot the morning after Thanksgiving. There were enough assorted work crews representing enough companies that we had to explain we were from Devon and were looking for Rocky Cowan, the company man who would show us around.

Cowan looks like he was destined from birth to be nicknamed "Rocky." In spite of all his industry experience, he still has all ten of his fingers. He took us into the trailer where he and several men in red Halliburton coveralls were preparing their next fracturing job, also called a fracturing stage. The drilling rig that had drilled this site just a few days ago made a bore hole that went down vertically then curved out horizontally. On a computer monitor, Cowan showed us a graph tracing the path of the bore hole. The hole was shaped like a boot, with a vertical shaft, a heel, and a toe. Into that bore hole the drilling crew had inserted steel pipe, known as a casing. The casing has cement on the outside preventing the gas that would flow through from communicating with the rock formation around it. There is also cement in the casing, until it reaches the shale sections. Then the crew must perforate a hole in the casing, opening it up to the rock formation. Wasting no time, the crew, working aboveground, injects the hydraulic mix into the hole, cracking that rock. The gas seeps through the hole, which is then quickly plugged so water and sand can't follow the gas back into the casing. Once all of the stages are complete the plugs are drilled out to allow gas to flow.

Before the fracturing crew set up, the drilling crew had inserted a wire line into the bore hole. The line contains instruments measuring the whereabouts of the perforating equipment and sends that information back to the crew in the trailer. The fracturing process starts at the toe of the bore hole and works its way back toward the heel, nearest the bore hole. Jay Ewing explained later that engineers determine where holes should be perforated and at what distances apart from one another.

At the time of my visit, Cowan and crew were tracking the location for their next perforation. Although they can accomplish several perforations, or "frack jobs" per day, we arrived when they were just in the preparation and planning stage. A good time to chat, but not a good time to experience the ground being split under our feet, so deep that only the computer screens in the trailer could bear witness.

Work here would be completed in just a few days, and then the production pipeline would be installed. This gas well would likely bubble along on its own, requiring minimal daily production maintenance, for many years to come. Instead of tall rigs, open reserve pits, and throbbing diesel engines, travelers in this area would be left with basically inoffensive tanks to look at, no worse looking than all the area's other visual detritus. Depending on how many wells were drilled from one well pad, an area might have several tanks clustered together. That is known as a tank battery, Cornell explained. The battery sits on a well pad at the end of a road, and that is what I noticed on my flight to Oklahoma back in June.

We left Cowan and the various crews to their work and set off to see what happened to the water used on the fracturing sites. As Fred Cornell drove us to the final stop on our tour, he and Leemaster described some of the issues involved with the startlingly large quantity of water used in hydraulic fracturing. Though it takes 3.5 million gallons of water to fracture each site, Leemaster said it was a small percentage of all the industrial water used in Texas. But when you consider how much industrial water the state of Texas uses, even a small percentage can add up to a lot of gallons. According to the U.S. Geological Survey, in 2005, industry in Texas withdrew somewhere between 1,500 and 3,200 million gallons per day of fresh surface and saline groundwater. It ranked third in the nation for industrial water withdrawals, behind Louisiana and Indiana.

We stopped at our final location, a place as peaceful as its name suggested: the Dove Hill recycling facility. There among a few trailers, outbuildings, and small water-treatment structures we easily located Anthony Smith, Devon completions foreman, and Brent Halldorson, chief operating officer with Fountain Quail Water Management out of Granbury, Texas. Smith has worked his way up through the company and learned about water recycling along the way. Halldorson is an engineer with a trimmed blonde beard and distinctly Canadian vowel sounds. His company is a subsidiary of the Canadian company Aqua Pure. It specializes in recycling "hard-to-treat industrial wastewater with modular mobile technology." It found an excellent client in Devon Energy.

I knew that the water supply for fracking in the Barnett came from regional reservoirs, municipal water supplies, and wells drilled into the Trinity River and Woodbine Aquifers that lie between the Barnett and the surface. The water left over after fracking was therefore unusable, and so was commonly pumped down disposal wells drilled far beneath the Barnett layer, removing it permanently from the water supply. Halldorson said the disposal required multiple truck trips over dusty roads to haul water from gas wells to disposal sites. This process was costly, but not as expensive as recycling was at one time. In some locations this water is simply diluted by being discharged into rivers and streams. If the dilution is not sufficient, the salty chemicals can affect drinking water, Halldorson said.

In a well-publicized instance of hydraulic fracturing unpleasantness, residents of Pavillion, Wyoming, population 160, located in a mature gas-field region of central Wyoming, began to complain that their well water had developed an unpleasant taste and smell. They attributed these changes to the hydraulic fracturing activity that had begun not long before they noticed the situation with the well water, around 2004. EnCana is the company doing most of the drilling in the area.

Luke Chavez is EPA Region 8 Superfund site assessment project manager, based in Denver. That is the agency overseeing the testing. He said that the EPA had responded to complaints by local residents and tested several classes of wells in March 2009. "When you get out there and see stuff floating around in the water and it smells and tastes funny and you don't want to drink it—it is just common sense." That's why since that time residents have been using bottled water for cooking and drinking, though Chavez said the EPA's tests "came back clean for the standard stuff" such as heavy metals and volatile organic compounds described in the Clean Water Act. In fact, of the thirty-seven wells tested, several came back showing no problem at all.

But the EPA does not have a complete list of chemicals used

in drilling and fracturing, because of the companies' unwillingness to disclose proprietary recipes. Further, EnCana says that because they hired out much of the work to subcontractors, they do not always know the exact chemicals used in drilling or fracturing. That makes identifying compounds in the well water problematic. Chavez explained that they try to diagnose the water problem the way a doctor would diagnose an illness: by collecting and considering all the data possible. "We looked at the agricultural practices and at the gas fields. We asked what they use in the drilling or fracturing process, what goes in the rig, what goes in the pits, and what present and past owners do."

Chavez said the EPA has two objectives: to identify whether there is an ongoing risk in the well water, and if so, to identify the source of that ongoing risk. The EPA has not gone into this investigation looking for proof that natural-gas drilling or hydraulic fracturing is linked to well-water problems in the area, although many local residents are pointing fingers in that direction. The EPA is simply looking for the answer. That answer was expected to be forthcoming after the agency finished examining data from its second round of testing, in 2010. They sampled twenty-one domestic wells on this occasion. The wells supplied drinking water in some cases, while others were for stock watering or irrigation. There were also two municipal water-supply wells in the group tested.

"When we get the data in we'll want to make sure the data are validated, the t's crossed, the i's dotted. Then we'll interpret the data, and then we'll present final results in May in Pavillion. That will be a very popular meeting," Chavez predicted.

When Brent Halldorson with Fountain Quail in Texas heard the story about Pavillion, he said he sympathized with the residents' worries about their well water. He wonders if some previous drilling activity in this mature gas field was the culprit, rather than hydraulic fracturing. He also characterizes as "politically motivated" some media coverage of the fracturing practice and the industry at large. "Water is a very emotional topic and often there is an element of truth to some stories that tends to get blown out of proportion and used to support scare tactics," Halldorson told me. "I feel that as I represent an environmental water treatment company, I can honestly tell you that shale gas fracturing is safe.

"I definitely do not blame the public for being concerned—there have been too many times in the past when people have said 'just trust me' and contaminated water has led to all sorts of terrible consequences."

However, in December 2011, EPA Region 8 released its draft report titled *Investigation of Well Water Contamination in Pavillion, Wyoming*. In it, they conclude that chemicals located in some of the deep wells they tested were not naturally occurring, but were of the sort commonly used in fracturing, such as the biocides and lubricants Fred Cornell had told me about. EPA suggested that those chemicals should not have leaked out of the gas wells and into the water supply, but that due to various environmental and man-made triggers, that's what likely happened.

According to the report:

> Detection of high concentrations of benzene, xylenes, gasoline range organics, diesel range organics, and total purgeable hydrocarbons in groundwater samples from shallow monitoring wells near pits indicates that pits are a source of shallow groundwater contamination in the area of investigation. When considered separately, pits represent potential source terms for localized groundwater plumes of unknown extent. When considered as whole they represent potential broader contamination of shallow ground water.

The EPA's draft report entered a public comment period so that various constituencies could review the research methodology and findings. What happens next will affect not just the situation in Wyoming, but in other areas where hydraulic fracturing occurs. The EPA report concludes by urging "greater transparency on chemical composition of hydraulic fracturing fluids, and greater emphasis on well construction and integrity requirements and testing. Implementation of these recommendations would decrease the likelihood of impact to groundwater and increase public confidence in the technology."

I made my visit to Dove Hill and my meeting with Brent Halldorson long before the EPA's Pavillion research was completed. He was quick to state his belief that Devon Energy is one of the good guys. Halldorson's views of what occurred in Pavillion might be reshaped after the EPA issues its final report about what happened in the EnCana gas fields. But when we met, he argued that companies like Fountain Quail would not be in business if not for "solid companies" like Devon Energy. He notes that Devon is not required by law to recycle the water used in hydraulic fracturing. It is common industry practice to re-inject contaminated water, known as frack flow-back, deep into the ground through disposal wells. Frack flow-back is water that is ejected initially out of the well due to high pressure. It makes up between 20 and 30 percent of water used in the process. The Ellenberger Formation is a ready holding area for this flow-back water. But Fountain Quail points out that re-injecting doesn't only require expensive trucking and permanently remove water from the environment's hydrological cycle. It also can destabilize the ground and aquifers. That's why they pursued boiling and distilling the water to recycle it.

Halldorson and Smith walked us around the facility to show us the water at various stages of the recycling process. Red tank trucks parked around the area contained contaminated water to be recycled. We saw water in tanks at various stages of apparent cleanness, rather like one would see at the average municipal water-treatment facility. They scooped water out of a tank into a beaker and dropped in a small amount of the polymers used to treat the water. Before our eyes the water in the beaker went from something like muddy creek water to water that looked more drinkable than it actually was.

The jewel in the crown of this facility is called the NOMAD 2000 evaporator system. NOMAD is able to process highly variable oilfield wastewater and return to the producer approximately 85 percent of it as distilled freshwater that can be reused. It can distill up to two thousand barrels per day of distilled water. And as its name implies, the NOMAD is small enough to be easily transported to the sites where industrial water recycling is needed.

"The distilled water being reused by Devon is more pure than river or drinking water due to the nature of the process," Halldorson told me. "The saltwater is boiled and the steam is recondensed as pure distilled water. About the only concern for the distilled water is that ammonia recondenses along with the water vapor for reuse—think fertilizer. This is not a concern; however, if we plan to discharge the water to a river or creek we would need to remove the ammonia to prevent algae growth, and so on. This can easily be done by polishing the water prior to discharge."

Typically, Fountain Quail does not discharge the water into rivers or creeks, but instead into a holding pond. There it waits until the next time Devon needs some of that 3.5 million gallons for a nearby fracturing job. Smith told our group that he'd been approached by locals wanting to use the pond as an area for duck hunting, since it attracted wildlife. He told them no.

It would have been easy to hang out in this quiet spot in the Texas countryside, but I had a long drive ahead of me back to Little Rock. So our group said good-bye to the small crew at Dove Hill and loaded back up into the Yukon. On our way back to Kusin's Convenience Store, Cornell drove us past a few wells in production. Some of the gas was no doubt headed to one of Texas's natural-gas power plants, to generate electricity.

While there are fewer natural-gas plants than there are plants that burn coal, there are still more than 730 of them across the country. Fifty are in Texas. Even environmental groups that are against the burning of fossil fuels, such as the National

Resources Defense Council (NRDC), concede that natural-gas power plants are cleaner and are more environmentally acceptable than coal plants. However, they are concerned about the process and drinking-water contamination occurring during fracturing. Rather than arguing for the shutdown of natural gas–fired power plants, they advocate for additional regulation of the drilling and fracturing process. The NRDC's Amy Mall says, "NRDC supports federal regulation of hydraulic fracturing under the Safe Drinking Water Act. We believe this is a sensible approach that would ensure a minimum federal floor of drinking-water protection in the more than thirty states where oil and gas production occurs."

Barnett Shale gas will be around for an estimated dozen or so years, at current levels of production. Chip Minty told me during my visit to Oklahoma City that the industry was only able to produce about 20 percent of the available shale gas there. "We're not smart enough to get 100 percent out," he said. But he predicted that advances could lead to the production of as much as 50 percent of the gas in place in the Barnett. And, the gas shale plays available nationwide should yield enough fuel for another one hundred years, at today's technology, he said.

The EIA's *Annual Energy Outlook 2011* report supports Minty's prediction about the increase of shale gas into the country's energy portfolio. The EIA indicates that in 2009, about 14 percent of natural gas produced was from a shale play. By 2020, they predict the percentage will rise to 45 percent. But they note considerable uncertainty about the amounts of recoverable shale gas in both developed and undeveloped areas. They also note that production rates can vary across individual shale plays and thus are hard to predict. "Environmental considerations, particularly in the area of water usage, lend additional uncertainty."

For the sake of argument, in a century or two the Barnett may be devoid of recoverable natural gas. As a student of history, to me a few hundred years does not seem very distant. I come from a place where things are slow to appear on the landscape and even slower to be reabsorbed. True, once production is finished in a well, the equipment is dismantled, the bore hole is plugged, and the land is reclaimed. But the remains of well pads and assorted industrial residue will be more enduring on the scene than the product they existed to create.

5

Homegrown Revolution

An Iowa Biomass Research Facility

JULY 1

The high temperature today was a relaxing 81 degrees, with a blanket-cuddling low of 49. This is peak summer in Laramie, with July bringing the warmest weather, the most robust afternoon thunderstorms, and the splashiest sunsets. With this nice weather, we put everything we have into Independence Day celebrations, when "Lar-amigos" pack into Washington Park to celebrate Freedom Has a Birthday, as it is known. After sheltering at home during the usual monsoonal afternoon thunderstorm the crowd heads to War Memorial Stadium on the campus of the University of Wyoming for a country-western concert and fireworks display. This event is the unofficial kickoff for Jubilee Days, a week of rodeos and street dances celebrating the day in 1890 when Wyoming joined the Union.

How fortunate we are that these two major causes for celebration occur in July. Summer is the payoff for living in a climate that keeps us in polar fleece much of the year. In fact, the widespread truism that there are two seasons—winter and the Fourth of July—might well have been spoken first here. I have personally witnessed snow in Laramie in July. It was 1993, my first full summer here after moving to Wyoming. My

Midwestern relatives were visiting at the time and also experienced the light wet snow spitting through dark low clouds. Through polite smiles they promised to visit again another summer. They never did. But snow in summer really is unusual, so we still set about planting flowers and vegetables, gardening as if growing season wasn't squeezed between Memorial and Labor Days. As if, as the poet Diane LeBlanc says, there was hope in Zone 4.

This summer, though, reminds me of that one in 1993. Although it hasn't snowed we have had much more rain than usual. In fact, although July started with reassuring warmth, by the Fourth the high was only in the low 60s, and the holiday country-western concert and fireworks show was rained out. We chatted with friends and neighbors across the fence to try to recall whether or not this is normal for this time of year. We tried to conjure up what life was like before the extreme drought conditions that gripped our area for the last several years. Did it really used to rain this much? Were summer days really so often this cool? Was there really no interval between summer evenings too chilly to sit out without a sweater and the arrival of mosquitoes that vex us back in?

Although the city's mosquito abaters were finally able to spray most of the biting critters into submission with Bti and permetherin, there was nothing they could do about the temperature. Luckily, twenty years ago my husband purchased a wood burning stove, but he had never used it. In this chilly midsummer he decided to drag it out of the garage. One night, as the evening temperatures lurked in the low 60s, we split up some firewood we'd stacked for use in the indoor fireplace. Also in the pile were the remains of our apple tree that no amount of coaxing, spraying, or praying would pull back from death's brink. We crumpled up newspaper, added kindling, and lit our first backyard summer fire. Finally, we were warm enough to sit outside with a snifter of brandy and a skewer of marshmallows, watching the sun set. Thank you, biomass fuel.

Basically, biomass fuel is anything renewable that was once alive that one burns for energy, excluding fossil fuels (dinosaurs aren't renewable). That means that biomass is either wood, waste, or alcohol fuels. Wood can include dead trees removed from forests, or urban materials like leftovers from the paper industry. Agricultural waste can include leftover crop material, or even methane gas produced from hog farms. When pioneers and early homesteaders threw buffalo chips on the fire, that fuel was waste from biomass. Lastly, alcohol fuels are mostly ethanol, frequently refined from corn but increasingly from other plants or grasses, known as "energy crops." Biomass is considered renewable because, even though wood, for example, is burned, more wood can be grown to take its place.

Alcohol fuels like ethanol are used mostly as a transportation fuel, an additive to gasoline that was touted as a way to lessen our dependence on foreign sources of oil. A controversy has played itself out in boardrooms, political debates, and farmers co-ops ever since ethanol became viable on a commercial scale. If too much corn is diverted from feeding people and livestock, what will happen if grocery prices skyrocket because everything from corn chips to beef becomes costlier? Some observers point to Brazil, where they grow sugarcane as an energy crop. That takes sugar out of production as a food product, and the world faces occasional shortages and attendant rising prices, as a result.

In addition to its use for producing ethanol, biomass fuel can be burned along with or in place of fuels like coal at a power plant to create electricity. Or even more efficiently, it can be digested by bacteria, which create methane gas that can also be used to turn turbines and create electricity. There are now about a hundred power plants running on biofuels putting electricity onto the nation's grid.

I had heard that biomass fuel was a promising source of energy because it is carbon neutral. This means that whatever carbon dioxide is released by its burning will be reabsorbed by the plants that take the place of the original biomass sources, assuming any do. I'd also heard that the fossil-fuel energy it took to use biomass on a commercial scale required more power than it could add to the grid. That struck me as a negative trait worth investigating. I wanted to understand how biomass materials could create electricity, power my car, and form the fabric of the clothes I wear. So I contacted Robert Cleaves, chair of the Biomass Energy Association based in Portland, Maine.

I asked him about what seemed to be a frenzy of interest in renewable energy in the early 1980s, followed by a surge of apathy. He explained the legislative effort, pushed by the Carter administration, to introduce conservation by the public and by utility companies. In 1978, Congress passed the Public Utility Regulatory Policy Act, or PURPA. It required utilities to purchase some of their power from independent producers, including producers of wind, solar, and biomass energy.

That led me to my second question, devised when I saw a map of where most biofuel power plants are located. The plants aren't concentrated in Midwestern farm states, where I would have guessed they'd be. According to Cleaves, after PURPA passed, "there were a lot of entrepreneurs wildcatting. They went to areas of the country that were obvious sources of fuel generation: woody biomass from forestry operations, such as New England and the Pacific Northwest, and agricultural nonwood material. These include sugar wastes found in Florida, rice hulls found in Louisiana, and orchard prunings in California," he said.

"You won't find them in Iowa because, while there is agricultural waste, it isn't the kind that is appropriate in combustion in typical boilers used in industry. In order to be viable, producers must have realizable sources of solid fuel," Cleaves said.

He then put me in touch with one of those producers, Mike Whiting, of Florida-based Decker Energy International. Florida became the home of Decker Energy when it was founded in 1982 "because it was a nice place to live," Whiting told me. Whiting is one of the founders of the company, which he owns with his family. Decker's stated business objective is to "develop, acquire, and own energy generating facilities that meet the needs of the modern electrical market by providing reliable, low-cost power with a keen sensitivity to the environment."

I spoke to Whiting from my cell phone on the upstairs deck at the Bear Tree Bar and Grill in Centennial, Wyoming, the first place I could acquire a signal after a summer morning hiking in the Snowy Range mountains. I told him about the blue, cloudless sky above the deck and the hummingbirds zinging past me into an aspen grove. I also mentioned how I'd spent my morning hike: scouting out dead trees I'd return to later to harvest for the woodstove and fireplace. First I'd need to obtain my firewood permit from the Forest Service.

Whiting sees a great future for electricity creation from biofuels. "It makes up about 1 percent of all electricity and about 3 percent of all energy in the U.S. That's not insignificant," he said. He compared biomass power to other forms of electricity such as solar and wind. "Other forms aren't as reliable. The consumer wants the lights to come on when they hit the switch. Biomass runs 24/7."

That's because biomass can be burned in a traditional baseload power plant as a substitute for other fuels. In addition to reliability, Whiting sees the opportunity for growth in biomass partly because of the pulp-and-paper industry. The waste from these manufacturing processes needs to wind up somewhere, so it might as well be burned in a biomass power plant, he said.

Decker Energy plants burn a variety of fuels. For example, their Grayling Generating Station in Michigan is a 37.2-megawatt plant that burns waste wood from local sawmills and the forest product industry. Their Craven County Wood Energy plant in North Carolina is a 48-megawatt plant that burns wood waste. Their Ridge Generating Station near Auburndale, Florida, is a 39.6-megawatt plant that burns urban wood waste, scrap tires, and landfill gas.

Whiting said the process of creating electricity at biomass power plants is the same as at most other types of plants. A fuel source is burned to boil water resulting in steam, which stimulates a turbine, which spins to create electricity that goes out onto the power grid. "We're basically like a giant tea kettle," Whiting said.

As with other types of power plants, there is sometimes local opposition to constructing a new plant. Whiting said Decker's plants are "pretty well received." I wondered if people concerned about the environmental impacts of a coal-fired plant, for example, would be more welcoming of a plant that burns materials that would otherwise end up in a landfill. Whiting said that in his experience most people don't care what you are doing as long as it is not in their backyard.

"There are right and wrong places to site biomass power plants," he said. "They are fine from a distance but if you are too close, the plant is a noisy eyesore." For starters, there are trucks arriving all day and night bringing in the fuel that will be burned. The company tries to locate plants in areas where there is already busy industry and major roads so trucks aren't running through neighborhoods or through the center of small towns. And more importantly, they want to site plants so they are near existing power transmission lines. Decker's plant sites are near established infrastructure, "not out in some remote area of the West like where you are in Wyoming," he said.

I took that last in the spirit in which it was intended, but I also was pleased I'd chosen to live in an area not overrun with industry, where a biomass plant would be just another noisy eyesore. The few scattered power plants we have here, all coal fired, are as surprising to the eye as a wolverine in the dining room.

According to the American Biomass Association and most observers, biomass power plants are an improvement over fossil-fuel plants. They contribute less to greenhouse gases than fossil-fired power plants. They efficiently combust the methane associated with power plants and eliminate it, emitting all the biomass carbon in the form of carbon dioxide, one of the greenhouse gases.

Yes, carbon dioxide—a gas to be avoided, according to what I'd learned on my power travels and other research. The EPA's Climate Change web page provides guidance on how to think about the hazards of this gas. First, it is emitted both naturally and through human activities like the burning of fossil fuels. "Natural" refers to the "carbon cycle," in which "billions of tons of atmospheric carbon dioxide are removed from the atmosphere by oceans and growing plants, also known as 'sinks,' and are emitted back into the atmosphere annually through natural processes also known as 'sources.'" According to the EPA it was during the Industrial Revolution that people intensified their burning of coal, oil, and gas, while also removing large swaths of forest; in 2005, global atmospheric concentrations of carbon dioxide were 35 percent higher than they were before the Industrial Revolution.

"When in balance, the total carbon dioxide emissions and removals from the entire carbon cycle are roughly equal." Assuming burning biomass is roughly carbon-neutral, and assuming that "sinks" such as trees will exist to offset carbon from the burning of wood, the pollution from burning biomass isn't as bad as burning other fuel sources. If that sounds like a good news–bad news compliment, it is. But the complexity of using biomass fuel puts it in good company, according to Nathanael Greene, the renewable energy coordinator for the National Resources Defense Council. The NRDC is a large environmental action group whose mission is to protect wildlife and wild places and to ensure a healthy environment for all life on earth. When Greene and I spoke, he understood my concern about forms of renewable energy being less benign than I had hoped.

"I haven't found an instance where there isn't a potential for a double-edged sword," he said about new developments in renewable energy. "If we use the right materials and technology, biomass can reduce pollution and a host of other environmental impacts. However if we aren't careful we can exacerbate the same environmental concerns that would make us want to use more of that form of renewable energy to begin with."

Greene speaks to Mike Whiting's points about using materials that would otherwise be headed for a landfill to create electricity in a power plant. Yet he offers a slightly different point of view. "From a greenhouse gas perspective, most biomass in the waste stream actually has a higher value used either in recycling or composting. Paper makes up a tremendous portion of the waste stream, but it is the easiest, most cost-effective material to recycle and avoids the need to go out and chop down trees and convert our forests into plantations" to replace that paper. He argues that setting up a recycling program is dramatically more environmentally beneficial than burning paper for power.

However, Greene does support burning contaminated wood in biomass power plants, if it means diverting it from landfills. Contaminated woods can come from construction and demolition projects embedded with nails or other building materials. It can also include the sort of treated lumber you and I might use to build a backyard deck. "Contaminated wood that ends up in landfills can lead to the most noxious toxic chemicals and water pollution in landfills," Greene said. He endorses burning these materials "if they are burned in a state-of-the-art facility, or gasified, or [handled by] any other technology that uses the best air-pollution control—that's actually a much safer way of disposing of these materials at the end of their life."

Greene explained why we need to understand the full picture of biomass, not just for power creation, but also for transportation fuel. "In the long run, there's only a finite amount of land and biomass that we can afford to divert for energy. You can't think about biopower without thinking about biofuels and understand what the cumulative demand on our landscape will be," he said. He explained that it is important to realize that we'll need far more biomass for energy than we'll ever be able to make. He emphasizes user conservation, but just as importantly, power plant and transportation efficiency, as the way to maximize not only renewables, but more traditional forms of energy as well.

"Energy efficiency is the fastest, cheapest, most effective way to reduce all sorts of the environmental impacts of energy production," he said. "We need to make sure that's always at the top of our hierarchy and the center of focus of our policies. The benefits of really strong energy efficiency and conservation go beyond just fossil fuels and directly into making it easier to integrate renewables into the mix." Greene points to the challenges associated with intermittent forms of power, such as wind and solar. If we have a grossly inefficient power transmission system, it will be very expensive to integrate those systems into it. He explained that if the system is efficient we can integrate less power into it—in other words, if we use the various fuel sources we have in a smarter way, we'll need less of it. And that would be good, because, as Greene said, "Even renewables have their environmental footprint."

I realized after speaking with Nathanael Greene that I wasn't going to be able to isolate my thinking about biomass to focus simply on electricity. I needed to learn more about the use of biomass for transportation fuel. Biomass fuel makes me think ethanol, which makes me think corn, which makes me think Iowa. So I arranged to visit Norman K. Olson, the Iowa Energy Center's BECON facility director. BECON stands for Biomass Energy Conversion Facility. It is operated by Iowa State University, and is located in the small town of Nevada, just outside of Ames.

The Iowa Energy Center is a research, education, and demonstration organization dedicated to improving Iowa's economy and environment by advancing the state's energy efficiency and renewable energy use. According to Olson, "The main objectives at BECON are to strengthen the rural economy and decrease U.S. dependence on imported petroleum by developing cost-effective methods of converting agricultural-based plant materials into value-added chemicals and fuels."

Olson is a registered professional engineer with a degree in mechanical engineering from Iowa State. He recalls that when he was in college, OPEC (the Organization of Petroleum Exporting Countries) reacted against pro-Israel U.S. policies and decided to cut oil production and place an embargo on what they shipped to this country. That caused what is today known as the 1973 oil crisis. "Crisis" is not a word Olson likes to pair with oil. When he thinks of U.S. dependence on sources of oil from countries that aren't necessarily our friends, he says he envisions an oil rig tossed on stormy seas, violently flaring fire. "The oil crisis must have been traumatic for me," he said.

At any rate, in college he found himself interested in coursework about solar and other forms of renewable energy. That interest and resulting academic emphasis led him to various professional positions including chief engineer for the Iowa Energy Policy Council, energy manager for the University of Iowa, and project manager for the Iowa Energy Center's Energy Resource Station.

I am not the first person who has contacted BECON lately to ask questions. Olson said he has noticed an uptick in interest over the last few years about biomass and other forms of renewable energy. More than eleven thousand people have seen his online slideshow about BECON or visited the facility, ranging from curious senior citizen groups to representatives from other nations wanting to know about the technology under development.

Olson greeted me in the BECON lobby on the morning of my visit. With trimmed gray hair ringing his head and a blue polo shirt the color of his eyes, he looked like he'd be equally at home driving a tractor or sipping chai at Iowa State's Campustown. First we sat in a large conference room and Olson fired up the laptop and projector. He launched the slideshow I'd viewed earlier on my home computer. It had made little sense to me then, lacking as it was in narration. He talked me through it this time and although I took notes, I realized I wasn't going to morph into a chemical engineer in the next few minutes. Short of that background, I could at least ask the basic questions most observers would have about projects at BECON.

In answer, he walked me around the facility where the research projects were underway. BECON is surrounded by cornfields in the town's business park. It houses a fully equipped analytical lab with a tub grinder, feed mill, steam boiler, thermal fluid unit, and a microturbine. The grain and storage processing units make it possible to take raw feedstock to finished commodities, Olson explained. And the biodiesel fuel produced there feeds a generator that provides some of the heat and electricity for the facility.

Some of the projects Olson showed me were being developed by private industry, while others represented research by Iowa State faculty and graduate students. Projects underway at the time were described to me as research on the industrial use of sweet sorghum, supercritical water gasification of biomass, supercritical fluid processing of biomass to chemicals and fuels, and evaluation of lignin/phenol derived from bio-oil production for use as an antioxidant in asphalt. Olson showed me around the projects, most of which were being attended to by young men engrossed in their work. And he helpfully defined the term *supercritical* for me, which refers to any substance at a temperature and pressure above its critical point, where distinct liquid and gas phases do not exist.

As we walked around, Olson told me that not too many

decades ago many goods were made from agricultural products. Biomass-based chemicals like cellulose, ethanol, methanol, and vegetable oils had been used since the 1800s to make paint, glue, adhesives, synthetic cloth, and other items. Since the 1930s the petroleum industry has learned to refine petrochemicals to create these items and has generally taken over the marketplace. Olson advocates for the development of biorefineries that would allow biomass products to reclaim that role and relieve our dependency on petroleum products for everyday goods. In one sense, that would be a way to relieve our dependency on oil altogether, rather than just foreign oil, the stuff we so often hear described as the chief threat to our nation's security.

BECON's researchers look for ways to improve the technology so that biomass can be converted into chemicals to make a variety of items, in addition to fuel. There are two main ways to convert biomass. Although before my first visit I pictured Bugs Bunny tossing an ear of corn into a Magic Oven and having it come out a sundress, this is not one of the two ways. No, the biomass conversion systems researched at BECON are a bit more complex. Olson explained that the biological conversion method uses bacteria to convert the biomass, and the thermal-chemical conversion method uses heat and chemicals.

Of these systems under research at BECON, Olson said he is least confident of the biodiesel project. He is concerned that there isn't enough vegetable oil to sustain this technique, in spite of do-it-yourselfers converting their vehicles to burn used French-fry oil. The one he is most hopeful about is alcohol fermentation. But the bottom line is that none of these techniques will work without commercial-scale application. "If you want to make a change out there, there has to be a business," he said.

Business is starting to pick up in the biofuel industry, in spite of Olson's concern about the quantity of available French-fry oil. Mark and Matthew Roberts run a California company called Springboard Biodiesel. They make machines that allow users to process their own biodiesel fuel in a small-scale but almost fully automated way.

"We sell equipment to people who mostly use collected waste cooking oil," Matt Roberts told me. "Most aren't doing it in order to wave the 'green' flag—they don't care about burning something clean. They want to burn something that saves them money." Their customer base is "anybody who feeds a lot of people," such as university sustainability managers. Another category of customer is people who have access to the leftover cooking grease after a lot of people are fed. Alachua County, Florida, which includes the city of Gainesville, is a Springboard Biodiesel customer. The county has long had a robust solid waste and recycling program, including multiple drop-off points and a wide variety of materials accepted for recycling.

Kurt Seaburg is the county's hazardous waste coordinator. He has a degree in environmental science and has been with the county department sixteen years. He told me that when the agency's director heard about a local demonstration project in which children used cooking oil to power a lawn tractor, he thought it would be right in line with the sorts of things the county was already doing to reduce its own carbon footprint. So the search was on to find a way to turn the cooking oil they were already collecting into biodiesel fuel. That's how they found Springboard Biodiesel.

Seaburg said the machine they purchased is simple to use and doesn't require too much hands-on work. They are able to make fifty gallons of biodiesel in a forty-eight-hour period. They use a blend of traditional diesel with 5 percent added biofuel, known as B-5. That blend goes into the county's off-road vehicles, such as front-end loaders.

The county is applying for an EPA grant to conduct an outreach program, educating residents about how the cooking oil they recycle is being used. Additionally, they are preparing to purchase a twenty-kilowatt generator that would power their 8,500-square-foot facility during daytime hours. The generator would run on 100 percent biodiesel, which Seaburg said is a little bit of a hurdle to achieve. That's because no generators made in the United States are currently "warranteed" to run on pure biodiesel, Seaburg said, although that isn't the case in other countries.

Seaburg figures they'll likely void the generator's warranty the minute they switch it on by using fuel not recommended by the manufacturer. He said they've run tests on their biodiesel fuel already, in line with standards set by the American Society for Testing and Materials, and feel confident the quality of what they are making will work fine in their generator. That'll enable them to be "off the grid" during the day, and use biodiesel as both a transportation fuel and as a way to generate electrical power on their own.

Matt Roberts is confident about biofuels and their future. "Look at how much diesel gets consumed by this country and the global economy, how many billions of gallons get used. Look at mandates from all these countries and how much they must use. Biodiesel isn't going away. The thing about biodiesel that is fabulous is that you can make it from so many different feedstocks. It can be made from a long list of vegetables, seed crops, rendered animal fat. A lot of these sources would be thrown away."

Another item some regard as waste is corn stover. That's the term for stalks and leaves left over in fields after corn is harvested. One organization interested in the possibilities of stover is Plains Justice. They provide legal resources "to help communities in the northern plains make the transition to a new energy future." They enlisted Iowa State University biofuels researcher Mark Mba Wright to study whether Iowa could replace much of the coal it burns in power plants with leftovers from the cornfields. They see the issue as having both environmental and economic importance. Wright's research shows that Iowa produces 68.3 million tons of corn stover a year, as well as 20.4 million tons of soybean residue. While some of the corn and soybean residues need to remain on the land to keep the soil healthy, much of it is readily available for fuel, according to Plains Justice.

"Burning corn stover in power plants would give farmers an income boost while helping to keep Iowa's air clean by reducing the amount of coal burned to produce electricity," according to Nicole Shalla of Plains Justice.

Organizations such as the World Resource Institute study the possibilities and pitfalls of using food products like corn for fuel. An economist there, Liz Marshall, explains that "as increasing world food prices heat up the food versus fuel debate, and scaling up corn production for ethanol use raises environmental concerns, increased attention has turned to the potential for second-generation ethanol technologies to free the domestic ethanol industry from its dependence on corn grain."

Whether corn stover will have the desired effect and replace corn grain is doubtful at the present time, according to Marshall. Under current financial and technical realities, farmers can't easily transition to methods negating the loss of corn stover, which keeps topsoil in place and helps to fertilize next year's crop. "Even moderate harvest of corn stover and other agricultural residues for use as an ethanol raw material, or feedstock, threatens to significantly increase erosion and emissions of greenhouse gases from the agricultural sector," according to Marshall.

Norm Olson at BECON is not a supporter of using corn itself for fuel. "I believe in food, first," he said. But he does support the use of other agricultural products including corn stover. At the conclusion of my tour we returned to the BECON lobby, where several stocks of sweet sorghum stood in a corner. This is one of Olson's favorite crop products currently being researched as a source for commercially available ethanol. Sweet sorghum yields about as much ethanol per bushel as corn, with another advantage: the stems are used for fuel, but the grain is still usable as food.

I asked Olson about the criticism I'd heard about biomass fuel: that traditional forms of electrical power need to be in place to conduct biomass conversion to fuel. In other words, somewhere, something is burned to create steam to turn the turbine that puts the electricity into the biomass plant. Olson acknowledges that this is the case, but notes that biomass conversion can be accomplished with any form of power, from coal and nuclear to solar or wind.

Robert Cleaves, of the Biomass Energy Association, argues that biomass is best used for electricity, rather than for transportation fuel such as ethanol or biodiesel. "On a BTU basis, dollar for dollar, biomass to energy makes more sense than biomass to fuel," he told me. But Olson at BECON favors producing transportation fuels from biomass and keeping the whole process confined to the native geographic region. "The first reason is that we import the majority of our transportation fuels and it is more beneficial to our economy to replace imported fuels than domestic fuels—it creates more local jobs," Olson explained.

"The second reason is that transportation fuels are generally higher in cost than the fuels used to produce electricity—mainly coal and natural gas in the U.S. It is generally easier to compete against these higher-priced transportation fuels," he added.

Fuel, transportation, corn stover, energy crops: Why aren't the people I ask about these aspects of biomass in agreement? Feeling befuddled, I paid a call to Vijay Sethi, vice president of Western Research Institute's Energy Production and Generation Business Division, based a few miles outside of Laramie. Sethi's expertise covers a broad range, including coal combustion and gasification, coal upgrading and efficient coal utilization, clean liquid fuel production from natural gas and other gas sources, and biomass as a fuel and for clean fuel production.

The day of my visit was one of those rare January afternoons with temperatures in the 20s but no wind. Several inches of snow covered the ground, but the sunshine and blue sky had tempted Sethi and two of his colleagues outside. When I tracked them down, they were driving a bucket of pink golf balls out toward the railroad tracks behind WRI's Advanced Technology Center. "Compression testing," Sethi explained.

The other thing he explained to me was about the contrary and opposing opinions I'd encountered. "They are all right, in their way," he said.

He explained to me that the easiest thing to do with biomass is to burn it. "We know how to burn things—humankind has been doing that for a long time." Burning biomass to create electricity is a pretty good idea, he said. It is also a smart idea to add biomass fuel to what is combusted in a traditional coal-fired plant. A plant might burn four parts coal and one part biomass, he said. The problem is that wood, for example, doesn't burn the same way that coal does. A process called *torrefaction* is a way of roasting the biomass, rather like the way coffee beans are roasted, to make them more compatible, he said. If consumers want "green" energy, they have a chance to pressure producers to add carbon-neutral sources to their power production. But consumers have to be willing to pay more for it, Sethi pointed out. "Somebody has to be able to make money" before entrepreneurs embrace various technologies, he said.

As to whether corn stover and other agricultural wastes are best left in the field to fertilize next year's crop, or used instead to be converted into ethanol, Sethi said the economics of the technologies involved have left that question without an answer, for now. "The government is throwing lots of money at this issue, so the technology could be getting closer."

Sethi is gratified to see the "second generation" of biomass technology move past using food crops for fuel. "I came to the U.S. from a third-world country. I would rather use corn to feed people instead of using it for fuel," Sethi said. "Waste is a step in the right direction."

Being immersed in these conversations got me seeing

everything from corn crops to mountaintops as so much fuel. I am not proud of mentally reducing natural landscapes to their industrial properties, but I live in a part of the country where it is easy to do so. In Wyoming I'm surrounded by natural gas, in evidence from all the drilling rigs poking up in much of the countryside. I could see the results of Powder River coal mining while driving through the Nebraska Sand Hills on my way to BECON. My route along Highway 2 paralleled that of the Burlington Northern and Santa Fe railway. Trains with four or more engines lugged 150 or so hopper cars, each loaded with Wyoming coal. Furthermore, wind farms are everywhere in Wyoming, hydroelectric dams, though modest, do add to the grid, and solar and geothermal are under development. Last but not least, I'm surrounded by trees, a fine source of biomass if there ever was one.

My home in Laramie is only about a dozen miles from the eastern section of the Medicine Bow–Routt National Forest. Farther into Wyoming are numerous forests, and to the south, in Colorado, still more. When mountain-region dwellers say "forest" we usually mean national forests managed by the U.S. Forest Service, places we escape to for recreation. National forests are usually, though not always, in mountain ranges and are operated as "multiple use" areas. That means hikers, campers, ATV riders, ranchers, loggers, hunters, and bird watchers have to find a way to pursue their competing activities in finite amounts of space.

Few people live in national forests unless their dwellings were homesteaded before the national forest boundaries were established. This is known as being "grandfathered in." Chief among the residents of these forests are the trees, which are fuel for fire in the natural order of things. The fires thin out the undergrowth, promote the success of various forest plants and animals, and help trees propagate through flinging their seeds. The forest service has tried several approaches to forest fire management, from complete suppression when fires start, to aggressive logging and undergrowth trimming to reduce the fuel source. They've also tried not doing anything, allowing fire to achieve its natural relationship to fuel. That doesn't go over well when human-fabricated structures are threatened or when large swaths of national parks go up in flames, as Yellowstone National Park did in 1988.

Since around 1997 a new menace has threatened the towering lodge pole pines and spruce trees of western forests and backyards alike. The instrument of this threat is the bark beetle. The forests have taken on a reddish-brown tint from all the trees killed through their infestation. There are numerous beetle species, including pine bark and mountain spruce, which affect forests just about everywhere they stand. But in the last few years their numbers have caused an epidemic in the forests of Colorado and Wyoming.

The beetles burrow into the bark of mature lodge pole pines and spruce to lay their eggs. The trees fight back with oozing resin but usually lose the battle because the beetles attack in such great numbers. The result is slow death not just to one tree, but to entire forests as the creatures move through it. The reason this infestation has affected trees in epidemic numbers has to do with the temperature. The beetles cannot endure prolonged, extreme cold. Just the kind of cold we have not seen much of here since the mid-1990s. So while it seems pretty chilly to me up in the forest in my polar fleece and down of a winter afternoon, the beetles are snug as a bug in a rug. Now the infestation in Colorado and Wyoming has reached at least two million acres. Researchers say that with global temperatures on the rise, we'll have to find another way to combat the beetles if we want to preserve these trees. There's a significant possibility that the towering lodge pole pines and spruce will be no more during our lifetimes. In their place the aspen might stage a comeback, returning to their earlier days of dominance in the forest.

The trees already dead or dying pose a threat for widespread and very serious forest fires. The forest service is acting aggressively to remove standing dead trees around campgrounds and along highways. In the national forest near my home, many campgrounds have been closed because standing dead trees could fall without warning on the people beneath them. So the forest service has cut down these dead trees and heaped them in large slash piles. They've given out free permits so that people can go gather as much as two cords of wood per household. Recently my husband and I grabbed the chain saw and trailer and headed to a forest campground near a drained fishing lake, and cut up dead trees to feed into our outdoor wood stove and indoor fireplace. While some larger communities that suffer thermal inversion have either banned or restricted the use of woodstoves and fireplaces, Laramie has not done so. It likely never will, because our brisk winds keep particulate from settling into valleys and causing polluting haze as occurs elsewhere.

Some communities have decided to treat these dead and dying trees as a source of renewable fuel on a somewhat larger scale. They can get rid of the fire danger and contribute to the percentage of power created through biomass. Reacting to the beetle epidemic, Colorado senator Mark Udall has joined with the U.S. Forest Service and the Denver Water Board to seek U.S. Department of Energy funding for a proposed biomass plant in his state, near the ski-resort town of Vail. The plant would be fueled by local lodge pole pines that have been killed from the bark beetle infestation. The developer is seeking thirty million dollars in federal funding for the project, which would be built on three acres of municipal property. One wrinkle in this plan is determining what will happen to the plant once the fuel source is depleted. It is hard to imagine convoys of trucks rumbling up from Denver delivering urban paper waste to be burned in the plant.

Nathanael Greene of the NRDC takes a longer view of the various forest policies that position dead trees as "excess," as opposed to seeing them as part of the life cycle of the forest. "We've done something through our forest fire suppression policy. Letting invasive species in has screwed up the forest systems. The simple fact that a forest is in transition cannot be justification for going in and chopping it down." However, he suggests that discrete supplies of downed trees such as the one near Vail might be just right for burning in a local biomass power plant. Wood is very heavy and so it doesn't make economic sense to haul it more than fifty or seventy-five miles, he said. On a small scale, though, such as to provide power to a school or community center, the Vail plant might make perfect sense.

BECON's home, Iowa, in contrast to my western home, has no national forests, but it has plenty of hardwood trees, especially in the eastern part of the state. Additionally, Iowa is the number one producer of corn in the United States. It averages 2.05 billion bushels per year. Most of the crop goes for feeding livestock; some of it is processed for food, cornstarch, and plastics, and of course, ethanol production. Iowa also produces soybeans, hay, oats, Christmas trees, fruits, honey, and grapes. Those grapes supply more than thirty wineries and one hundred vineyards in Iowa.

I did a little sight-seeing as part of my BECON trip, heading east to Dubuque, a city that is making great strides toward both revitalization and sustainability. While there I visited the tasting room of Stone Cliff Winery, housed in a former brewery overlooking the Mississippi River. It and numerous other wineries are part of the Upper Mississippi River Valley viticulture region. This is the biggest wine region in the country, encompassing nearly thirty thousand square miles of Iowa, Illinois, Wisconsin, and Minnesota. Compare that to the 759 square miles of the more famous Napa Valley in California. That's a lot of acres devoted to a worthy cause, in my estimation.

The bartender at the Stone Cliff tasting room and I were

discussing her favorite and least favorite thing about living in Dubuque. I'd ordered the wine sampler—several selections from the wine list for five dollars, together with the souvenir glass. The bartender poured me sip-size samples and described each one. Between sips of Red Dog and Riesling, I munched chocolate chips and oyster crackers that had been decanted into wine glasses and placed on the bar.

She told me that what she liked least about Dubuque was the frequent "fish fly" hatches rising off the Mississippi River in the summer months. Also called mayflies, they seem attracted to her white house paint and leave behind quite the mess, she said. But her favorite thing about her native Dubuque was "just about everything else," from the weather to the people to the river itself. Then she remarked about what a beautiful area I live in and recalled a recent trip she'd taken through neighboring Colorado. "There is so much unused land there," she said, referring to the high arid plains of eastern Colorado. "Just think if they would plant crops on all that land. There would never be a food shortage again."

She was speaking of efficiency from the point of view of a person nurtured in a fecund land. When I see those same vast grass and sage prairies of Colorado and Wyoming, I realize that I live in a landscape whose purpose, if I must use that word, is openness. At least that's the way I see it. Others see the land as prime for the oil and gas industry, which drills deep into the earth to put fuel into our gas tanks and to heat our homes. Still others see the land as prime grazing for cattle, and indeed, ranching has both protected and exploited this landscape for more than a century. Before the homesteaders and barbed wire arrived, Native Americans used these lands for hunting buffalo and establishing home camps as the seasons changed. All of those land uses are fully practical, too.

If there could be one emblem for this combination of efficiency and beauty, to me it would be the humble firefly. We don't have them in the high country of the West, but to a child of the Midwest "lightning bugs" are the essence of summer. I'm drawn to Midwestern travel during the summer, hoping to catch a glimpse of them rising in unison from the grass just before dark, power packs on their back. Indeed, I saw such a spectacle in Ames the night before my BECON tour. I was crossing the expansive lawn of the National Veterinary Laboratory near my motel when hundreds of the flying light bulbs lifted from the tall dewy grass. I'm told that all that flashing is about mating. But maybe there is also a lesson there about using what you have and having what you need.

As one moves about in Iowa, one sees very few acres of land not produced, not turned into something useful in a practical sense, be it a farm field, a parking lot, or a groomed suburban lawn. There are many lovely state parks providing access to a fishing stream, or a place to park the RV in an improved campsite on a reservoir with several dozen other likeminded travelers. To its enormous credit, the state has started buying up much of the limestone bluff lands along the stretch of the Mississippi River in northern Iowa to hold as nature preserves where the landscape will not be "improved" by development. There is also an effort at cropland conservation, stewardship encouraging farmers to nurture plants other than corn, and animals other than livestock.

Although I appreciate and value the stewardship practiced by Western ranchers who hold large tracts mostly of land unbroken and available for wildlife, others disagree with me. Some people celebrate the breaking up of the large ranches into subdivisions so that more people can witness antelope playing and the sun setting each night behind the purple mountains. I have to admit there is something efficient about accessing natural beauty just by looking outside one's window. Efficiency of energy and resources is also on the minds of the Union of Concerned Scientists when they think of things once living now used for fuel. They note that biomass contains less energy per pound than fossil fuels. "This means that raw biomass typically can't be cost-effectively shipped more than about fifty miles before it is converted into fuel or energy. The advantage of this is that local, rural communities—and perhaps even individual farms—will be able to design energy systems that are self-sufficient, sustainable, and perfectly adapted to their own needs."

Just like the firefly. We should be so adept.

Journey a Little Way into the Earth

A Utah Geothermal Plant

AUGUST 1

In what has become an unnerving trend, August started out chilly in Laramie. The high on this day was 56 degrees, with rain, and the low was 39. I call this a trend because most months this summer have not only started cool but remained cool. I call it unnerving because with a summer like this, who needs winter? I'm not one of those who sit with binoculars trained at the mountain peaks waiting for the first glimpse of snow. I don't ski downhill and I cross-country ski mainly as a way to feel fresh air and sunlight on my face. I sometimes wonder what I am doing living in this permafrost place. Then I remind myself that there are also places of nine-month summers, suffocating humidity, poison ivy, and insects that could eat Cincinnati.

I celebrated my seventeen-year Wyoming anniversary by preparing for my trip to the Blundell Geothermal Plant outside of Milford, Utah. I sat on the backyard deck reading up on geothermal energy, warmed by my outdoor woodstove. I wrapped up in jeans and sweatshirt watching the hummingbirds, which arrived late this year. Normally they come and go with the delphinium blossoms, which is to say they arrive in mid-June and depart by late August. This year I'd convinced myself they weren't coming at all, in spite of the presence of six-foot-tall light-purple delphiniums and the birds' other favorite food source, the six-foot-tall pink hollyhocks. Most likely the chilly weather held them off but finally they arrived in late July, the broadtail and rufous varieties chattering and chasing each other off telephone lines and crabapple branches.

Planning my trip to Blundell got me thinking about Jules Verne's *Journey to the Center of the Earth*. I'd read a poor translation from French as a teen and developed a dislike for it. So I decided to rent the 1959 movie starring James Mason and, of all people, Pat Boone. What reason the filmmakers could have had to take a harrowing story about a dangerous scientific expedition and rewrite it to accommodate a white-patent-leather-shoed crooner is beyond me. But that's what they did, casting Boone as the young scientist narrator, Alex (from Axel, in the original), and throwing in red-headed Arlene Dahl as a completely new character, apparently so that a woman explorer could travel to the center of the earth, too. Then there's longhaired Hans and the Eider duck, portrayed by a white goose, but please, let's not think about it. Let's just say that watching the film prompted me to locate a better translation from the Project Gutenberg Consortia Center's World Public Library Collection.

In Verne's world, people can plausibly visit, if not the center of the earth, at least a giant sea half the way there. Monomaniacal Professor Lindenbrock, the young Axel, and Icelandic guide Hans penetrate the earth by descending an inactive volcano in Iceland. Eventually they are expelled along with great globs of lava through an active volcano, which lands them somewhere in Italy. In between they explore geological ages visible through their lava-hewn trail beneath the earth's crust. Before it is over they meet up with creatures from an antediluvian age and even find the physical remains of ancient man.

The story is fantasy, not real. But Verne's use of a scientist narrator—a young man charged with keeping the journey log—creates a sense of a real expedition. It reminded me a bit of the journals of Lewis and Clark, describing their almost-as-fantastic trip up the Missouri River, over mountain ranges, and ultimately to the Pacific Ocean. I only wish Lewis and Clark had been as consistently detailed as Verne in their description of events.

My own voyage to the Blundell plant in southwestern Utah took me through country no less daunting today than it must have been to early European explorers of this area. Lewis and Clark did not go this far south, but John C. Fremont and John Wesley Powell did. For the most part they had the sense to travel this country by water, not attempting to cross the endless ranges of mountains and mesas and desert that comprise southern Utah.

No James Mason or white geese were along for the ride with me, but I was happy to have the company of my husband, Ron, our Schipperke dog, and an eleven-hour audiobook of a Janet Evanovich mystery. Fortunately, our path was significantly easier to trace than the one followed by the early western explorers. We drove along Interstate 70 from a point in Colorado to the interstate's western terminus at Cove Fort, Utah. As someone who grew up along the oldest section of I-70 in Missouri, I never before considered where that highway began or ended, or even that it did. Call it provincialism or a profound lack of imagination, but I suspect I am not alone in taking for granted where things come from before they get to me or where they go when they depart. Things like electricity, to name an example.

Traveling west to the source of that 2,200-mile strip of asphalt made me feel I was descending into the ancient structures of the earth. Utah's I-70 takes the traveler along the longest stretch of interstate with no services, where exits are for viewing scenery or for testing the brakes of potentially runaway trucks. At a point about thirty miles west of Green

River, Utah, is the San Rafael Swell, a giant dome-shaped anticline of sandstone, shale, and limestone. Infrequent but powerful flash floods have eroded the sedimentary rocks into valleys, canyons, gorges, mesas, and buttes—the very sort of scenery most of us picture when the state license plate commands us to think "Utah!" But until the stretch of I-70 through this area was completed in 1970, very few people made the journey.

Like Lindenbrock's team dead-ending into walls of granite in their underground mazes, we smacked into I-15 from the end of I-70 about twenty miles north of Beaver, Utah, population 6,162. We spent the night there and in the morning headed out to Milford, the town the plant lists as its place of residence.

The Blundell plant itself is not in Milford, population 1,437. Instead, it is about fifteen miles northeast of town. We followed directions e-mailed to me by Garth Larsen, plant manager in charge of operations. I thought I'd be able to throw the directions away as we got closer, thinking I'd use the tall cooling towers and transmission lines that I expected would guide us in. But none of those things was visible. Instead, we drove along a gravel road through Bureau of Land Management (BLM) desert, past a wind farm under development, and over some railroad tracks. The thin traffic became nonexistent save for a water truck dampening the dusty gravel. I could not see the plant—only the Mineral Mountains, with Granite Peak rising 9,771 feet above sea level in the bright morning sun.

We were a few hundred yards from the plant when the road curved toward the gated entrance and I could finally see the structure. The cooling towers were not particularly tall and only one power line led away from the plant toward town. The main buildings were painted BLM Desert Tan, allowing them to blend in with their rocky surroundings. The entrance gate was locked. I could see only a few vehicles in the parking area and no one out moving around. I'd been instructed by Garth Larsen to dial the phone number printed on the phone box and whoever answered would let me let inside. I dialed the number. I got the answering machine.

I waited a few more minutes, listening to the screech of a raptor soaring in the thermals overhead. I dialed again, and much to my relief was greeted and let through a swinging gate into the facility. It wasn't long before Rene Andrews, Blundell's operations supervisor, came outside to greet me. He explained that Rene is spelled like the French name but not pronounced that way. I dedicated this fact to memory by realizing Rene rhymed with "green." Tanned and muscled, Andrews looks like he's taken good advantage of the outdoor lifestyle that dominates Utah. He chatted with Ron through our open truck window about area features he could visit while I toured the plant. The Mineral Mountains are a big favorite of rock hounds and climbers alike. Andrews gave directions to an obsidian dike in the Ranch Canyon, a little distance from the plant. Closer to the plant is visible evidence of the geothermal features that dictated the placement of the Blundell plant.

The site lies on the Roosevelt Hot Springs Known Geothermal Resource Area (KGRA). There were once hot springs here, popular with settlers, miners, and cattlemen who needed a place to wash up and relax after long periods on the trail. The surface hot springs were dried up by 1966 but on occasion steam still rises from the ground nearby. While there is no evidence left of a brothel that Andrews told us stood here in the nineteenth century, the structure of a bath house and pool once filled with geothermal water is still visible.

Andrews told my husband to watch his step and to keep an eye on our dog. After all, these hills are full of dangers such as rattlesnakes, spiders, and scorpions, he said, not to mention noxious fumes potentially emanating from the ground. I felt like I wasn't the only one who should be issued safety equipment as my sandal-wearing husband and our rodent-hunter dog motored out, the secure gate closing behind them. Hoping I hadn't said good-bye to them for the last time, I followed

Andrews inside the turbine-generator building and up the stairs to the conference room to commence my tour.

Andrews gave me a little more history of the area and told me about the plant. Blundell is a 34-megawatt facility that serves about 25,240 residential customers. The Roosevelt Hot Springs geothermal field ranges from 1,250 feet to 7,300 feet below the surface. It is eight miles long and six miles wide. Its water temperature is around 500 degrees Fahrenheit with a pressure of 500 pounds per square inch. This thermal reservoir is relatively near the surface of the earth thanks to the Negro Mag and Opal Dome faults, according to geologists who study the area.

Long ago earthquakes broke up the Tertiary granite and Precambrian metamorphic rock in the area to allow the magma-heated water to seep into the resulting faults. The Tertiary geologic period spans from 65 million to 2.588 million years ago, from the time the dinosaurs went extinct to the start of the most recent ice age. The Precambrian is said not to be a period but a super eon. It started when the earth was formed 4,500 million years ago and ended with the evolution of hard-shell animals 542 million years ago. The Mineral Mountains intruded themselves, in geology lingo, into the Precambrian rocks about 25 million years ago. Volcanoes in the area erupted, spewing a rock called rhyolite starting about 8 million years ago. They stopped spewing a mere 500,000 years ago, leaving rhyolite domes in the Mineral Mountains. This range forms the third border of the Roosevelt Hot Springs.

Andrews explained how the Roosevelt Hot Springs were discovered to be hot. A group of locals attempting to drill a water well in 1967 found hot water at a depth of 80 feet. Since that wasn't what they were after, they plugged that well and moved the rig over 300 feet. They drilled again, this time to a depth of 165 feet, and found hot water that flashed to steam with a whoosh when it reached the surface. The well was eventually deepened to 265 feet, and there drillers discovered a mixture of water and steam. That well was eventually plugged and abandoned, but it is known as the "discovery well" for the geothermal field.

The Phillips Petroleum Company began exploration of the area in 1972, meaning they drilled wells to learn more about the scope and depth of the geothermal feature. In 1980 they entered an agreement with Utah Power and Light Company to allow power sales. In 1982 Utah Power began construction on Blundell Unit 1. Blundell was the first commercial geothermal plant outside of California, where geothermal features associated with faults are more plentiful. The plant was named for a former president of Utah Power. It is now owned by PacifiCorp, which is a subsidiary of MidAmerican Energy Holdings Company. MEHC describes itself as a global leader in the production of energy from diversified fuel sources including geothermal, natural gas, hydroelectric, nuclear, coal, and wind.

Blundell Unit 1 is known as a single-flash system; it came online in 1984. It works through production wells that bring the high-pressure, heated water to the surface. Wellhead separators flash the geothermal fluid into liquid and vapor phases. The vapor phase, what most of us call steam, is collected from the production wells and directed into the power plant at temperatures of between 350 and 400 degrees Fahrenheit with pressure up to 109 pounds per square inch. The steam turns a turbine in the ordinary way of power plants. The turbine fires the generator, which creates the electricity. The voltage output is stepped up for transmission over the electrical lines that tie into the transmission system in Milford.

Formerly, the liquid geothermal brine was channeled back into the reservoir through gravity-fed injection wells, Andrews explained. Then Blundell Unit 2 entered service in 2007, using the spent geothermal fluid from Unit 1 as a heat source. Blundell Unit 3, a dual-flash system south of Units 1 and 2, when finished, should bring the overall net capacity of the Blundell complex to 60 megawatts. Until then, the plant's net capacity is 23 megawatts, which equals the energy that would be produced by burning roughly three hundred thousand barrels of oil annually, Andrews told me.

Through a large south-facing window in the conference room, Andrews pointed out one of the production wells where the water and vapor come up out of the earth, and the separator. Andrews said that when a well is drilled into the reservoir, the reservoir's pressure forces geothermal liquid up the well casing into the separator. Because of the separator's pressure control, about 18 percent of the liquid flashes to steam in the separator tank. The steam leaves in a large pipe and heads into Blundell 1 from the primary steam inlet.

From the window we could also see out onto the tabletop-flat desert floor atop the geothermal field. The sage and other vegetation was scorched from a wildfire that started on July 6, 2007. The sprawling lightning-caused fire burned through sage brush, cheat grass, and pinion juniper. Two motorists died in accidents on busy I-15 caused by the thick smoke. The fire burned nearly up to the front gate of the plant. It didn't destroy the plant due to the dedicated effort of firefighters, Andrews said.

Another noteworthy feature of the geothermal field is that the actual ground above it is warmer than surrounding areas. Winter snow, which Andrews said can commonly drift waist-high, drives wildlife and livestock to the warm, green field to graze. When the snow melts, or when rain falls, the water seeps through the ground back into the superhot underground reservoir.

We left the conference room, donned our safety equipment, and headed to see the electricity being generated with a little help from conditions deep inside the earth. Perhaps I had too much Jules Verne coloring my expectations, but I was disappointed to look through the floor and, instead of seeing five thousand feet into the fire and brimstone of the earth's crust, I saw instead some insulated pipes. But once I began to understand what was happening inside those pipes, the exotic quality of geothermal power came back to me.

The steam coming through the pipes that I could see through the floor vent came from a pocket in the earth's crust five thousand feet below me. That's almost as deep as Denver—dubbed the "Mile-High City"—is high. The earth's crust varies from 5 to 25 miles deep, so the drill holes for this plant just crack the surface, so to speak. A well would have to be drilled about 1,800 miles to get through the middle layer, called the mantle. Then to get to the earth's outer core, it would need to go another 1,380 miles. To hit the inner core, another 780 miles.

In spite of Verne's Professor Lindenbrock's theories about a cool center of the earth, the temperature at the core is about 9,000 degrees Fahrenheit. That fictional professor never got that far on his own journey. In fact, humans digging holes for other purposes have reached depths of only 12,800 feet. That's the deepest mine in the world, where miners search for gold in the Witwatersrand region of South Africa. In North America, the deepest hard-rock mine is in Ontario, Canada, where they search for copper and zinc ore at about 8,880 feet, with the shaft going another 800 feet deeper.

Andrews himself worked briefly in an underground coal mine near Price, Utah, where the temperature in the mine always stayed in the high 50s. He said this experience taught him that he preferred to work aboveground. Utah is no stranger to mining disasters, but an accident about 150 miles away from where Andrews once worked is still fresh in people's minds. The Crandall Creek Mine collapsed on August 6, 2007, a month to the day after the wildfire came near the Blundell plant. The collapse killed six coal miners, whose bodies still remain in the mountain. A second cave-in ten days later killed three rescuers and injured another six. Another miner who would have been in the mine with the six initial casualties had left the site and hadn't yet returned at the time of the collapse. His survivor's guilt led him to commit suicide a year later, according to the Federal Mine Safety and Health Administration. As a result, that miner's family is also covered by the settlement from the mine owner and related companies.

Although Rene Andrews is out of underground mining, he is a lifer when it comes to working in the power industry. Like many people in the field, he worked his way up the ladder, over a career of more than thirty years, learning one job and then getting promoted into a position of greater responsibility. That has meant relocating more than once with his wife and children as he has moved to various power plants around the West. Like many Blundell employees, he lives south of the plant in Cedar City, Utah, which means a daily round-trip commute of about one hundred miles.

Andrews drove me around Blundell in a white company van. The whole facility felt about the size of a modest nine-hole golf course. While he drove, he told me stories about motorcycle riding through Montana with his son, and scuba diving with his wife. As he pointed out the sights, Andrews told me about a still-unmarried twenty-seven-year-old son whose marital status is a bit alarming to that young man's parents. He seemed comfortable being on two mental tracks at once. "My family is the most important thing to me," he said. "This career has made it so I can put my family first."

The Blundell plant employs twenty-three full-time employees. These include a maintenance crew of five, and three administrators, including Andrews, Garth Larsen, and Steve Austin, whom I also met along my tour. The employees are all men except for one woman who is an administrative assistant and has authority over the warehouse. One of the men, Korte Young, was working in the control room on the day of my visit. He has been there since the early days of power generation at Blundell. Like in most control rooms at power plants, the work appears to an outsider about as exciting as playing checkers against yourself in a mirror. To watch dials, switches, and lights to make sure everything is working within "operating parameters" around the plant could only be exciting, I would think, when they aren't.

As did Andrews, most control room operators work their way up through the ranks and transfer around to various plants. There are plenty of schools for this kind of work. Students can study practical technologies like power plant, process plant, electrical power, electrical transmission, and nuclear technologies. After two years of study they can graduate with an associate degree. Then they are eligible for hire into the sorts of jobs that the old-timers worked their way into, Young told me.

Andrews and I said our good-byes to Young and some of the maintenance workers gathered around the turbine-generator building and hopped back into the van. Resuming our circuit around the facility, Andrews pointed out the cooling tower, where air cools the condensed geothermal steam into water. Large pipes carry the condensed steam to the top of the tower, and the resulting water flows along a redwood deck and falls into a basin at the bottom of the tower. During this process some of the water evaporates, puffing visible steam out of the top of the tower.

The greatest expense for a geothermal plant is the drilling of very deep wells. There are variables in the heat of the water and other problems connected to the noxious hydrogen sulfide gas in the geothermal resource. And since one cannot know for sure whether a selected site will prove adequate for its purpose, drilling can be an extremely risky proposition. Millions of dollars can be spent drilling just one well, which may not pan out. In contrast, to discover whether a site is suitable for wind development, one need only erect and monitor an inexpensive meteorological tower to determine wind velocity. If the wind there is not appropriate, one takes the tower down and totes it to another ridgeline, to put it simply. But assistance for geothermal power development must continue, Andrews told me, because it can fill an important role in the nation's energy portfolio. "No one is pulling the meter off their house. People want power," he said.

Technology is being developed in Germany and elsewhere that could cut drilling costs in half. It still doesn't help if the well that is drilled is not suitable. But once an adequate well site has been located, instead of drilling separate wells for pumping the water out and for pumping it back in to be reheated by the magma, they are using one well for both steps of the process.

Power sources such as wind will be here as long as we have the sun providing the energy. But unlike solar and wind, geothermal plants provide baseload power. Blundell can operate twenty-four hours a day, and Andrews said its efficiency is close to 40 percent. The efficiency of a plant is the percentage of the total energy content of a power plant's fuel that is converted into electricity. And in terms of availability, the plant is available to produce power about 95 percent of the time. Wind, of course, is intermittent, as Andrews reminded me more than once.

That water can be put back into the earth to be reheated is what makes geothermal energy attractive, says the Geothermal Energy Association (GEA). The GEA explains that geothermal is considered a renewable resource because "the heat emanating from the interior of the earth is essentially limitless. The heat continuously flowing from the earth's interior, which travels primarily by conduction, is estimated to be equivalent to 42 million megawatts of power, and is expected to remain so for billions of years to come, ensuring an inexhaustible supply of energy."

The GEA also touts the clean quality of geothermal power. "Unlike fossil-fuel power plants, no smoke is emitted from geothermal power plants, because no burning takes place; only steam is emitted from geothermal facilities," they explain. Some chemicals do emanate from the geothermal liquid and vapor. "Emissions of nitrous oxide, hydrogen sulfide, sulfur dioxide, particulate matter, and carbon dioxide are extremely low, especially when compared to fossil-fuel emissions."

Geothermal resources are located all over the world. One typically thinks of Iceland first when thinking of geothermal power. At least, that's what Verne's Professor Lindenbrock thought of when the subject came up. The five geothermal power plants in Iceland produce electricity and hot water for heating, and there are plenty of geothermal pools for soaking and swimming. With so many volcanoes and other geological features, Iceland is able to provide more than 20 percent of its electric power from geothermal resources. Almost all the rest there comes from hydro power, with fossil fuels contributing less than 1 percent. Like the firefly, Icelanders use what they have.

But geothermal power doesn't have to be confined to areas where volcanoes rise, where hot springs bubble, or where earthquake faults lurk. Many individuals around the country heat their residences with hot water from the ground, using geothermal heat pumps. Anyone whose home is built on Planet Earth could in theory drill a well and use the earth's warmth to heat the air in their home and even the water. One doesn't need to access water hot enough to make steam and create electricity, only hot enough to be warmer than the ambient air, according to David J. C. MacKay, author of *Sustainable Energy Without the Hot Air*. He explains how the heat travels to a house in a system of looped pipes and returns it back into the ground, much like the system at Blundell. In the summer, home geothermal heat pumps pull heat out of the house and put it back into the earth. These systems are pricey and so is the cost of drilling the initial well. However, once a heat pump is installed it requires very little maintenance and results in low utility bills. The homeowner can also feel good about not burning anything to heat the home.

Relatives on my husband's side have a house so heated, in Indiana. My brother-in-law, of hardy German stock, loves the heating system. But my late sister-in-law, Janet, she of the 5 percent body fat, was always chilly. However, Janet might have been chilly if she had had a rocking chair on the very precipice of the geothermal resource itself. Try convincing her that the earth's heat is a constant source of energy that is essentially inexhaustible, as long as rainwater and snowmelt continue to feed the underground thermal aquifers. "Great. Now turn it up," Janet might have replied.

But that family and millions of others have had a consciousness-raising experience when faced with rising utility bills over the last several years. Homes were once built with not a lot of attention paid to insulation. After all, fuel prices were pretty affordable, up until the energy crises of the 1970s. Now people have been asking all sorts of questions as they build new homes, as my relatives did, or attempt to retrofit existing homes with energy-efficient items, like we are doing.

John Krigger and Chris Dorsi have written a book that I and other home-improvers value. Their *Homeowner's Handbook to Energy Efficiency* first invites readers to understand how they and their homes are complicit in adding to the carbon content of the atmosphere. Readers are taught to understand their electric bills and think about what they are buying with that monthly check to the utility company. Water heating is tackled next, showing readers ways to figure out how much hot water they really need. The heating and cooling section provides tips on using ceiling fans, programmable thermostats, and the like. The authors talk about exterior landscaping, and finding and sealing air leaks against the whistling wind. One chapter helps homeowners understand whether or not their home would be a good candidate for photovoltaic cells. Another gives useful tips for keeping mold and mildew out of the house.

All of these suggestions push the concept of energy conservation. The authors note that the most energy-efficient homes have several things in common. They have building shells that are airtight and well insulated; they have small heating and cooling systems; they have windows that make smart use of the sun's rays; their appliances and lighting are as energy efficient as possible; and they use solar power whenever possible.

Applying these principals to my own house, I can see areas where improvement could be made but others where achieving better energy efficiency or adding home-scale renewable energy would not be practical or cost effective. For example, our wood construction home was built in 1958. We've added a great deal of insulation to the attic and walls, which essentially had none when we moved in. We replaced the furnace in 2000 when we bought the home because inspectors at the time of purchase discovered the furnace was defective. It was a wonder the old gentleman living there hadn't been asphyxiated by carbon monoxide. In addition, there was a wall heater, like what one might find in an old mobile home, heating an addition to the house not connected to the main furnace. We never had the gas company test it, but we suspected just from its appearance that it was not only inefficient, it was dangerous. We replaced it with a modern stand-alone gas stove, the kind with the shiny paint, decorative doors, and fake flames. It came with a remote control, which we've never figured out how to use, but it also works on a programmable thermostat. That is one of two programmable thermostats in our home; the other controls the main furnace and brings the heat up from 62 to 70 each morning (then back to 68 when we're safely out of the shower).

The house has six sets of huge picture windows facing to the north and west. In the summer I'm glad the high-country sunshine isn't beating directly into the house, and in winter I'm happy for the sunshine, indirect as it is. We've replaced all the appliances since we've lived here, but it certainly seems that advances in efficiency outpace our ability to afford replacements. I know I could purchase a clothes washer that uses less water and less energy than the one I bought a few years ago. But my sense of not wasting a perfectly good appliance outpaces my desire to conserve.

In our home of 1,850 square feet, we operate four ceiling fans. The ones that include lights have rheostats, allowing us to dim or brighten the illumination as needed. We don't want or need air conditioning here at 7,200 feet but the ceiling fans produce nice cool air in summer, and move the heat off the ceilings and closer to our feet in winter.

After the burnt-orange 1970s wall-to-wall carpet became too worn to make excuses for, we replaced it with cork flooring. Cork, like bamboo, is touted as a renewable resource because it is harvested from living trees, not from trees that are harvested for lumber. It is much warmer underfoot than tile, for example. Softer too—I can point to kitchen dishware that I've accidentally bounced off the cork floor, and which have survived the impact nicely.

As far as landscaping goes, we have benefitted from two freakish accidents of weather. One year, an especially heavy June snowstorm took down trees in full leaf all over town. In our yard, it split the eastern blue spruce the previous owner had planted, right down the middle. The tree was healthy but the grass in that section of the yard was always brown and brittle from excessive sunlight and terrible topsoil. So we removed the tree, and the grass, and put in a xeriscape garden full of drought- and cold-tolerant plants like yucca and Russian sage.

The second freakish weather incident occurred when a tornado came through Laramie. I lived in the Midwest for thirty-two years, but the only time I've directly suffered tornado damage was here in Laramie, where tornados are rare. This tornado damaged the roof of the house enough that the insurance company paid for a replacement. We selected a forty-year Energy Star roof. It cost a bit more than what the insurance company would pay, but the $1,500 tax credit in place at the time made up for most of the difference.

Alternative energy sources such as photovoltaic aren't

practical for homeowners in all areas of the country, including my part. Neither are geothermal heat pumps serving individual homeowners, such as my husband's family uses. Of the 8 percent of electricity in the United States that comes from renewable resources, geothermal makes up only 5 percent, according to the Energy Information Administration (EIA). As Brian Hayes writes in *Infrastructure: A Field Guide to the Industrial Landscape*: "Why bother burning coal to make steam when you can just drill a hole in the ground and let the steam come whistling out? The only trouble is it works only in a few places in the world."

One problem is that not every location has relatively shallow geothermal resources near tectonic plates like the Roosevelt Hot Springs. Developing geothermal power is capital-intensive for reasons Rene Andrews noted. In addition, they are typically located in remote areas, as Blundell's rural location illustrates. Developing power is one thing: getting it onto the power grid is another.

The struggle in Wyoming to build transmission lines to move its wind power through remote and scenic locations illustrates the type of challenge faced by developers of other sorts of renewable energy, in this case, geothermal. Federal and state governments have developed various policies to increase the use of renewable energy, according to the EIA. The Renewable Electricity Production Tax Credit is a federal incentive that has encouraged an eight-fold increase of wind energy capacity since 2001. An expiration date for those credits was built into the law that established them. The tax credit for wind expires at the end of 2012. The tax credit for incremental hydro (adding hydro power to existing dams), wave and tidal energy, geothermal, municipal solid waste, and bioenergy was extended until the end of 2013.

Environmental organizations as well as governmental groups and the renewable energy industry are working to convince Congress to further extend these credits. Geothermal projects typically require between four and eight years to complete, according to the GEA. That means that many geothermal projects under development will not be completed by the current tax-credit expiration deadline, which will undermine future industry growth. They note that since 2005, the U.S. geothermal market has grown from 2,737 megawatts of installed baseload capacity in 2005 to 3,102 megawatts in 2010.

In order to advocate for this cause, GEA Executive Director Karl Gawel submitted a statement to the Select Revenue Measures and Oversight congressional subcommittees in fall 2011. It frames the challenges of developing geothermal energy in terms of other types of renewables. "We understand that a principal reason for providing solar projects the 2016 deadline was the long lead times expected for concentrated solar power projects," according to Gawel. "We believe that geothermal projects, with considerably longer lead times than currently faced by solar projects, warrant a comparable time frame."

Baseload geothermal provides reliability, Gawel said in his statement, as well as economic stimulation and job creation. "Geothermal resources in the U.S. remain largely untapped, because of the high risk of finding and proving geothermal resources. With continued incentives for investment in new power projects we will capitalize on new technologies which could make significant new geothermal energy production a reality in the U.S. and sustain U.S. leadership in the world geothermal market."

In addition to federal incentives, the EIA reports that many states have renewable portfolio standards that require electricity providers to generate or acquire a percentage of generation from renewable sources. However, states have "escape clauses" in these standards if renewable generation exceeds a cost threshold. Some states have delayed compliance and others lack enforcement procedures. Some states have built Renewable Energy Certificates (or credits) into their Renewable Portfolio Standards. This allows electricity providers to sell Renewable

Energy Certificates and use their proceeds for renewable projects. GEA reports that up to 3,959.7 megawatts of new geothermal power plant capacity is currently under development in the United States. Those states with projects currently under consideration or development include Alaska, Arizona, California, Colorado, Florida, Hawaii, Idaho, Nevada, New Mexico, Oregon, Utah, Washington, and Wyoming.

While capital costs, lead time, and geographic isolation are sometimes daunting issues, geothermal energy has many benefits. For example, geothermal plants are scalable—a small geothermal plant can be built to provide power for a small community, whereas a nuclear plant for that purpose with today's designs would be overkill. The GEA says geothermal plants require one to eight acres of land per megawatt created, versus five to ten acres per megawatt for nuclear operations, and nineteen acres per megawatt for coal power plants. Geothermal isn't a thirsty form of power requiring vast reservoirs to cool its apparatus. Plants use twenty liters of fresh water per megawatt-hour versus more than a thousand liters per megawatt-hour for nuclear, coal, or oil, according to the GEA. In terms of safety, geothermal advocates point out that we aren't importing the earth's heat from countries that are politically unstable and we aren't held hostage by pricing consortiums. The plants are not likely to blow up or melt down. They don't burn fuels that depend on removing chunks of mountain tops or digging up the ground beneath our feet.

Andrews, my initial plant contact Garth Larsen, and I were finishing up our conversation in the conference room about geothermal's advantages when we saw my husband's vehicle approaching the electronically controlled gates. As we went outside to greet him and hear about his adventures, Andrews pointed out the landscaping work done by a plant employee to formally beautify the already lovely desert terrain. Although water at the plant isn't potable, and so fresh water must be delivered in large plastic bottles, the supply there works just fine for the Russian sage and black-eyed Susan at the base of the transmission tower.

Professor Lindenbrock and company in *Journey to the Center of the Earth* were fascinated by the idea of a trip into the earth with little awareness of the danger. At least not on the part of the professor and the unflappable Hans—Axel the narrator was a nervous wreck. Most of us would not willingly climb around the tops of active volcanoes, even between eruptions, and the professor had no intention of doing so, either. That's why he and the crew followed fictitious Arne Saknussemm's path down the throat of Snaefells in Iceland, a volcano extinct for five hundred years. Only because of circumstances beyond their control did they exit through a "Strombolian" type of eruption involving intermittent bursts of expanding gas.

Just before that amazing deliverance, Verne's professor speaks the sort of lines that drive the best of science fiction literature. His sentiment could be applied to today's urgent question of energy use and conservation. "The situation is virtually hopeless, but there exists a possibility of salvation, and it is that possibility which I am examining. If we may die at any moment, we may also at any moment be saved. Let us then be prepared to seize upon the smallest advantage."

7

Water, Water, Everywhere

A Kentucky Hydropower Plant

OCTOBER 1

Just a few days back from my September swing through the Southwest, I had hopped into the nineteen-seat "Vomit Comet" in Laramie to catch a connecting flight out of Denver. From there I flew to Little Rock, Arkansas, where I rented a little red sport sedan that looked racier than it really was. By the time I was cruising along an Arkansas back road, a storm was moving into the Rocky Mountains. When I awoke the next day in Brinkley, Arkansas, and stepped outside into the nurturing dew of a southern morning, my friends in Laramie were being greeted by six inches of snow that had fallen in the night. That was September 21, the last day of summer.

It isn't unusual for Laramie to say "so long" to summer in that fashion. We watch all our carefully cultivated flowers and vegetables get smacked down by this annual September storm. Several inches of snow are accompanied by its riding companion, killing frost. Then a few days later the skies clear, the days warm, and the plants droop lifelessly to the ground.

Lest readers get the wrong idea about my town, the average high temperature on October 1 in Laramie is 63 degrees, and the low 39. Most years, October skies are blue, aspen leaves yellow, foothills golden brown. After our traditional end of summer blizzard it usually stays pleasant until Halloween. That's when we get our second traditional blizzard, and normally that's all she wrote for seeing the surface of any streets until April.

This year's Halloween weather will be anyone's guess. In the meantime we mark another Laramie tradition on October 2: the annual rubber-duck race down a short stretch of the Laramie River through town. Hundreds of rubber ducks are set loose to float a short course down the river. Locals buy tickets from the Rotarians and turn out to see which duck will bust out of the pack, the first to dodge eddies and snags, the first to be netted out and declared the winner. My duck didn't win. But watching that flotilla got me considering the Laramie River, and how a rubber duck or anything else has little chance of floating from headwaters to mouth without getting gobbled by a dam.

The Laramie River starts out in the Rocky Mountains of northern Colorado. It flows past the Medicine Bow Mountains in Wyoming and heads for the plains where it is impounded by a dam to form the Wheatland Reservoir. Some water is diverted there for irrigation. What is left of the river flows back out of the reservoir, where it receives some other smaller rivers, and ultimately dumps its watery load into the North Platte River after a trip of 216 miles.

The North Platte also begins in Colorado and flows into Wyoming. It meanders through the basins and ranges of central Wyoming and emerges from the mountains near Casper. It is impounded by several dams along the way to create reservoirs before the Laramie River joins what is left of the North Platte near the town of Fort Laramie. From there it heads east into Nebraska where it flows parallel to the South Platte.

The South Platte originates in the Rockies southwest of Denver. It too is impounded by a dam to form Chatfield Reservoir, a major source of drinking water for Denver. Then the river fights its way east, is joined by numerous small streams, and beats it to Nebraska. At the town of North Platte, the two directional Plattes form the Platte River. Off it goes across the state, draining into the Missouri River, south of Omaha. In its glory days the Platte River used to be a mile wide, but now is about the width of a four-lane highway due to so much water impounded in dams and removed for irrigation in Nebraska. The Missouri is lucky to get much sustenance at all from the Rocky Mountain drainages.

The story of this river system is not uncommon. There are an estimated seventy-five thousand dams in the United States, blocking six hundred thousand miles of river, according to the U.S. Geological Survey. That's about 17 percent of rivers in the nation that do not flow freely. Some of the dams are primarily for hydroelectric power and do not include storage. Some dams exist primarily for navigation, pooling water to allow barge traffic to travel through a system of locks. Some are there mainly for flood control, keeping raging rivers in check by stabilizing the amount of water that flows through at any one time. Others are mostly for irrigation, collecting water in a holding tank, so to speak, and then allowing it to flow through irrigation canals to nourish crops. Water managers must balance the needs of farmers with the needs of fish and other species in danger of being left high and dry if too much water leaves their habitat.

The control of nature has its advantages, depending on whom you talk to. Recreationalists love to boat on the reservoirs and camp along the placid shorelines. Locals rest more securely knowing their area river will behave more predictably as a lake than it did as a river susceptible to drought and flood. Farmers appreciate having water available for irrigation. And almost everyone loves hitting a light switch and having the power come on. Of the 8 percent of the nation's energy that comes from renewable sources, 34 percent comes from hydroelectric power plants, according to the Energy Information Administration.

But like most things that have fans, dams also have detractors. For example, there are people who once lived alongside a river and have lost their property under a manmade lake, uninhabitable forever, thanks to a dam. And many environmental groups advocate for the free flow of water and fish that are affected by dams. In some places around the country, especially the Pacific Northwest, grassroots organizations are fighting to dismantle existing dams. A group called the Save our Wild Salmon (SOS—the W doesn't appear in their acronym) has the endorsement of various conservation organizations, commercial and sports fishing associations, businesses, river groups, and taxpayer advocates. Their mission is to "restore self-sustaining, abundant, and harvestable populations of wild salmon and steelhead to rivers, streams and oceans of the Pacific Salmon states."

The group focuses its efforts on the Columbia and Snake River Basins. The Snake originates in southwestern Montana then flows south through Wyoming, northwest into Idaho, and at the Oregon border enters Hells Canyon, part of the Columbia River Plateau. The Snake winds into eastern Washington and after seven impoundments between Hells Canyon and the Columbia, makes its way to the confluence with the Columbia near Kennewick. Together with the waters of many other Columbia and Snake tributaries, the waters head for the Pacific Ocean where they mingle near present-day Astoria, Oregon. That's across the river from the spot where in November 1805, William Clark wrote in his journal "Ocian in View! O! The Joy!" His Corps of Discovery was actually looking at the Columbia Estuary, not the Pacific Ocean, but after two years struggling upstream over barely navigable rivers, then over the continental divide and downstream along the Snake and Columbia, their excitement is understandable. SOS seems nostalgic for the era of Lewis and Clark's travels, when sixteen million wild salmon and steelhead swam upstream to spawn in the lakes and rivers of the Columbia River Basin each year.

The goal of SOS is to return to that condition by removing four dams on the lower Snake River in eastern Washington, upstream of the Columbia. This is part of a larger solution to rehabilitate fish populations and to promote free-flowing rivers "as a vital economic engine for local communities." Their argument has merit, according to John Harrison of the Northwest Power and Conservation Council. However, "many river communities depend on the dammed river, with its irrigation, barge transportation, and recreation as a vital economic engine. It works both ways," he told me.

The four dams targeted by SOS for removal are the Ice Harbor, Lower Monumental, Little Goose, and Lower Granite. The Army Corps of Engineers built the dams between 1962 and 1975 for navigation, originally, with hydroelectric power as an added feature, followed by recreation. They are known as "run-of-river dams." This type creates only a small reservoir backing up behind each dam. A run-of-river dam creates power by being placed in a location where the river elevation drops gradually. This drop in elevation causes the water to flow fast through a tube called a penstock. The penstock directs the water flow through an adjacent power plant's turbines, which turn to generate electricity, which heads out through nearby transmission lines. Meantime, the water rejoins the flow in the main river channel, none the worse for wear.

Paradoxically, people fighting for the restoration of free-flowing rivers for the health of the ecosystem have another issue to consider. The hydroelectric power produced by those dams creates nonpolluting electricity: nothing is burned so no greenhouse gases are emitted. Nevertheless, some people have concerns about related issues and would rather see these dams gone, for a wide set of reasons both environmental and political.

Thanks in no small part to the Grand Coulee Dam on the Columbia River, the state of Washington is the nation's leader in producing hydroelectric power. More than 30 percent of the nation's hydroelectric power is generated in that state. Washington produces more than twice the amount of hydroelectric power as does California, which takes second in this category. The dams under consideration for removal have a combined maximum generating capacity of 3,000 megawatts. Since they only use the water available, not water that has been stored, they aren't always capable of creating that amount of power. When the water is low due to seasonal variance, less water flows past the turbines, resulting in lower electricity generation. Therefore, SOS says the dams aren't generating enough power to make harming the fish worth the cost.

Statistics about the effect of the dams on fish are kept by organizations such as the Fish Passage Center (FPC). It coordinates the Smolt Monitoring Program in the Columbia River system. A smolt is a young salmon that has not yet swum out to sea, which is what it does well before making its storied upstream run to spawn. Data from the monitoring program provide information for federal, state, and tribal groups about fish passage in the Federal Columbia River Power System. The FPC provides data on salmon, steelhead, bull trout, and lamprey throughout the Columbia River hydroelectric system. Salmon managers from state and local fisheries, as well as local tribes, use these data to develop policy about fish passage and migration. The FPC has concluded that goals for adult fish and smolt numbers have not yet been met, in spite of greater fish numbers released by hatcheries.

According to Harrison, several of the strongest runs of adult salmon in the Columbia River Basin in recent years have been in the heavily dammed Snake. "Dam opponents like SOS will say that's because of massive hatchery releases, which mask the effects of the dams—more fish released means more fish that return from the ocean to spawn. But dam advocates will say that passage survival of juveniles has improved vastly, and this, combined with favorable ocean conditions, is what's really driving the larger run sizes."

The true impact on fish notwithstanding, dams do contribute to the overall power demands of the Pacific Northwest and its hydroelectric customers. The dams won't simply be dismantled and be done with. Much of that power would have to be replaced, somehow, even with conservation measures in place. SOS argues that improved efficiency coupled with renewables like wind or biofuels, which don't harm fish, would do the job. One wonders then about the potential harm to wildlife and the viewscape, but nagging downsides are present even with renewables. An unanswered question is whether or not that power could really be replaced efficiently.

The Northwest Power and Conservation Council, where John Harrison works, is the official power-planning agency for the Northwest. They advocate investments in 5,800 megawatts of new conservation over the next twenty years, plus more renewable energy. According to Harrison, "If the power plan goals are met, it would mean that fewer new generating plants would have to be built to meet increasing demand for power." Harrison believes the direct replacement of hydro with other forms of power would not work and speaks to the question of efficiency.

"Wind turbines only generate power when the wind blows, and it doesn't blow all the time. Thus, a 100-megawatt wind turbine only generates on average, 33 megawatts of power, and that is the maximum average output. It can be as low as 15. Thus, 3,000 megawatts of wind-power capacity is worth about 1,000 megawatts or less of actual generation."

Harrison said the council neither ties the output of the four Snake dams to the potential wind-power capacity nor suggests that wind output could replace the dams' output. "The SOS folks say that wind power and conservation could replace the output of the four Snakes, but it is not that simple. We come close in our power plan, which calls for massive amounts of conservation, efficiency, and new wind, but we don't go so far as to say that could replace the four Snakes." He explains that the four Snakes supply baseload power that would have to be replaced with a reliable power source if it were lost because of dam removal:

> Ironically, if the dams were removed, the replacement power most likely would be from new natural gas–fired plants, which emit greenhouse gases. Our analyses have shown that dam removal would increase carbon dioxide emissions by four million tons annually because of the new natural gas plants required to make up a portion of the lost baseload power. Also ironically, hydropower is the primary source of backup power for the intermittent wind power in the Northwest. So the four Snakes contribute backup power to ensure that lots of new wind plants can be built.

The bottom line, according to Harrison, is that "for the purpose of planning our Columbia River Basin Fish and Wildlife Program, and particularly the hydropower system portion of the program, the council assumes that, in the near term, the breaching of any dams in the Columbia and Snake Rivers will not occur." The council takes a look at these issues regularly, at least every five years. "If, within that five-year period, the status of the lower Snake River dams or any other major component of the Columbia River hydropower system has changed, the council can take that into account as part of the review process," he said.

There are all sorts and styles of dams, in addition to the run-of-river sort in the Pacific Northwest. Some are made of mud, some of steel, some of timber. Some are built by humans, some by beaver. Not all exist for electricity production. That's obviously a rather recent purpose in the millennia-old history of dam building, dating back to a dam built in Jordan, in 3,000 BC. Dams that are built for hydropower work one of several ways. For example, the pumped-storage type uses two reservoirs, one at a much higher elevation than the other. At night and on weekends, when power demand drops, water that has been collected in the lower reservoir is piped up into the higher one for storage. When power is needed, that water is sent whooshing back to the lower reservoir, moving electricity-generating turbines as it passes.

The more common form of hydroelectric dam collects water in a single reservoir. The water flows down a river bearing fish, sediment, the occasional rubber duck, and in the case of the lower Columbia, ocean-going freighters. The water backs up against the dam and starts to pool. This pool is the reservoir. When the intake gates of the dam are opened the water flows through penstock tubes. Water spins the turbine blades, which are connected to the generator to create electricity. The water is then free to go on its way, but the creatures and sediment that were once in the river may not make it through the dam. In the case of the ocean-going freighters and barges, that's where the locks come in, to help them move through the obstacle that is a dam.

Some environmental groups caution against hydropower for environmental reasons, as the Snake River example illustrates. Others, such as the Environmental Literacy Council, simply speak to diminishing water supplies as a reason to curtail developing hydropower. "While the use of water to produce electricity is an attractive alternative to fossil fuels, the technology must still overcome obstacles related to space requirements, building costs, environmental impacts, and the displacement of people. However, within the U.S., possible locations for new hydropower projects are beginning to diminish."

It was becoming clear to me after this preliminary research that commercial hydropower is not a monolithic entity. Although it doesn't burn anything to produce energy, there is a certain extractive quality to it, like coal or gas mined for fuel. When we speak of rivers we really speak of watersheds, of heads, and of mouths, of systems in ways I've never heard applied to wind or solar. Water flowing unimpeded along a river has a nature unlike water that has been forced through a dam. The act of slowing it down, backing it up, and releasing it on a schedule not of its own making is to make water another substance all together. I wanted to more clearly understand how dams for hydropower work and how to think about what I'd always considered a renewable, green form of energy. To do that, I needed to see some big dams firsthand. I'd need to research more than one river, firsthand. I thought back to my sophomore U.S. history class at Bishop Miege High School, in the mid-1970s. Mr. McBride was teaching us about the Tennessee Valley Authority, or TVA. Mr. McBride's cheeks would flush red when he got worked up. And he got worked up talking about the TVA, originally created by an act of Congress in 1933 to help poor farmers during the Depression by managing the feast-or-famine Tennessee River and bringing them into the modern world with electrical power.

As he described the TVA's mission shift in World War II to providing electricity for aluminum producers like Alcoa to aid in the war effort, I recall my teacher's cheeks getting redder. Then as he discussed TVA's appropriation of land and damming of tributaries beyond the Tennessee, we students could see his politics showing. Then he told us about the contemporary battle over Tellico Dam, which achieved national prominence as an environmental and social issue. In spite of snail darters, Cherokee cultural artifacts, Early American battle sites, and white settlements dating to the early nineteenth century, the Little Tennessee River was impounded and the valley where those things dwelled flooded. The Tellico fight had been a forty-year battle, but in 1979 Tellico Lake finished filling, impounding the final stretch of free water on the river. Those of us who were paying attention in class must have felt a little flush in our cheeks, too.

I was thinking of Mr. McBride when I pointed my red rental car to Paducah in extreme southwestern Kentucky, to just a few miles from where the Tennessee flows into the Ohio on their way to the Mississippi. I reached the confluence of the two rivers in Paducah the afternoon before my scheduled visit to Kentucky Lake Dam. I sat on a bench watching the river flow and a barge being pushed along the channel by a tugboat. Barge watching is a hobby along big rivers, like train watching is along dry land. On this humid afternoon a couple I took to be retired sat on a shady bench with a marine radio. They were monitoring the communication of the boat captain and lock crew. Over the crackle of the radio, I heard someone use the navigational term "mark." Nothing new to anyone used to river traffic but a thrill to me to be reminded of the writer Samuel Clemens. He is said to have taken his pen name Mark Twain based on his time piloting steamers up and down the Mississippi. My family often visited his boyhood home of Hannibal, Missouri, on the banks of that river, when I was a child. That is partly what fed into my obsession with the Big River. That and the fact that my Uncle Jay had floated a stretch of the river south of St. Louis all alone as a young man in the 1930s, when the river was considerably wilder than it is today. That story ranks high in family lore.

I stayed at Kentucky Dam Village State Resort Park that night, just outside of Paducah. Kentucky Reservoir has a surface area of 160,300 acres and about two thousand miles of shoreline, so I picked the accommodations that seemed closest to the dam, on paper, anyway. When I arrived I was pleased to see the dam and the power lines in the distance from the balcony of my room. But I couldn't get there from the hotel. The road over the dam to the power plant and visitor center was under construction, marked with large "Road Closed" signs. I tend to be high-strung driving in unfamiliar territory for appointments when I don't know where I'm going, so I got an early start in the morning to reconnoiter the spot. For several hours before my scheduled 1:00 p.m. meeting, I drove up and down the Land Between the Lakes, along Lake Barkley, which is the other lake that the land is between, and across a blue bridge over the Cumberland River. I stopped for directions to Kentucky Dam several times and received conflicting advice. I got myself so turned around that when I saw a dam I actually thought I had succeeded in finding the place.

Relaxing for an hour on a bench overlooking this perfectly nice but completely wrong dam, I munched Chex Mix and watched people fish. I headed over early to the visitor center only because a man on a riding mower was determined to cut the grass under my bench. Thank goodness it was groundskeeping day. It was only after checking in with a helpful young lady at the visitor center that I learned I was still lost. I was a few miles off course and had found Barkley Dam, a project of the Corps of Engineers. Darn.

Thanks to the young lady I managed to make cell-phone contact with Jeff Ring. He is the head technician at Kentucky Dam, who was to be my guide. From his view out the visitor center's panoramic window, Ring could see my approach through light rain that was getting heavier. He talked me down the closed road across violent dips and loose gravel as I picked my way through a carnival of road construction equipment. "Are you driving a little red car?" he asked "Thank God, you see me!" I'm afraid I squealed. "Just aim toward those power lines and head downhill." Hydro dam in view. O! The Joy!

If Ring thought I was an idiot, he didn't let on. Perhaps it was his southern manners, or maybe he understood why I took the words "Road Closed" at face value. At any rate, he greeted me with a smile and held the door open as I dashed in through the rain. I found myself in a room about the size of a small-town bank lobby. Several kiosks allowed unattended visitors to hit a button and learn the history of TVA or how the various hydroelectric power plants in the system operate. One display showed the Raccoon Mountain pumped-storage plant near Chattanooga. It showed how water is pumped from Nickajack Reservoir at the base of the mountain, up to the reservoir. Then it is released when needed, through a tunnel drilled into the center of the mountain, like a snow cone leaking from its tip.

Ring directed my attention to a diagram of how the Kentucky

Lake Dam produces power. This plant uses a Kaplan blade turbine and has five generating units producing 199 megawatts. The dam is 206 feet high and 8,422 feet long. Ring was explaining how water enters the intake and flows into the penstock when there was a mighty clap of thunder and the popping sound of something electrical tripping. He left me alone momentarily while he scurried off to be sure that nothing important had blown.

While I waited I noticed one wall was basically a window allowing viewers to observe the plant's control room. There were rows of computer monitors showing graphs of river flow and generator efficiency, the sorts of things I'd come to expect in power plants. But the room was unattended. When Ring returned he told me that no one works in the control room any more. "It is all run from Chattanooga," he explained.

Ring himself used to be a control room operator before he became head technician, or "head tech." He's been in this business around thirty years, and he told me that others at Kentucky Dam have been around there even longer. The fact that no one at the plant was really running the show didn't appear to bother him. "We know what to do or who to call" if there are issues at the dam to be resolved.

He explained to me that the Tennessee River Valley contains the nation's fifth-largest river system. The TVA's goal is to "reduce flood damage, produce power, maintain navigation, provide recreational opportunities, and protect water quality in the 41,000-square-mile watershed." The TVA system includes twenty-nine hydroelectric plants. The hydropower system includes forty-nine dams in seven states from Knoxville, Tennessee, to Paducah and on to the tributary systems to the North Carolina border. TVA also operates three nuclear plants, eleven coal-burning power plants, and an assortment of solar- and wind-energy sites, along with a methane gas facility.

Ring walked me onto a balcony at the visitor center where we could stand out of the rain and view the gray thunder of the Tennessee River sluicing through the dam like beer froth through a harmonica. In spite of the noisy water, rainfall, and construction sounds, a great blue heron stood placidly on the riverbank. While we looked at the heron and the heron looked at fish, Ring told me about the TVA's priorities for the river. "River scheduling and navigation come first. Then electricity. But above all, human life and safety. If someone's safety was threatened, the other things wouldn't matter."

From our observation point on the balcony, Ring pointed out a new lock under construction that would help with navigation. The Tennessee River drops 513 feet in elevation from the time it leaves Knoxville until it joins the Ohio. The TVA built a system of nine dams along that stretch, allowing boats and barges to move along the river through a "staircase" of controlled current.

The area around Kentucky Dam had seen a great deal of rainfall the last few days, this day included. Ring told me he had been in conversation with the folks at Chattanooga about the amount of extra rainwater flowing over the spillway instead of being diverted through the penstock and past the turbines. The water in the reservoir had not quite reached its capacity, he said, yet "Chattanooga" was requiring him to let water flow free, since they were worried water could overflow the reservoir. "We're spilling without making power. Water is free, especially when it rains. We're losing money by doing that." But "Chattanooga" is the boss. And so is Knoxville, where the TVA is headquartered.

Anyone can stop by a TVA visitor center and see displays about the history of TVA and operation of a dam. It seems no one can tour a TVA dam's actual power plant. I tried for months to get permission and learned simply that "since 9/11" tours are not allowed. TVA says that as a government agency, they do not have the financial resources for the security required to let people into the power plants. Pamela M. Wilcoxon, senior management adviser for hydro production at TVA in

Knoxville, seemed sympathetic to my request but in the end was only able to arrange that my wanderings through the visitor center at Kentucky Dam be chaperoned. That's why my only experience of the workings at this plant was from videos created for tourists. Jeff Ring was very helpful and generous with his time when I visited. But sometimes it is nice when a local takes you under his wing.

I found my local in the form of John Harnon at the Kentucky Lake Bait and Tackle Store. The combination convenience store and gas station was a few miles away from my room at Kentucky Dam Village State Resort Park. The night before my tour I thought it might be nice to sip an adult beverage from my balcony while watching the lake softening with the colors of sunset. So I headed to the bait shop and walked expectantly to the cooler in the back of the store, but didn't find the hoped-for provisions. I approached the man behind the counter and asked where the beer was. He just shook his head with the air of someone who wished he'd had a lottery ticket for every time he'd been asked that question. "Dry county," he said with more empathy than I would have mustered in his place.

Travel writer Pico Iyer says we travel not just to get answers but to ask better questions. Harnon, the man behind the counter, gave me an earful of local information I never would have thought to ask about. He told me stories I'd best not repeat about moonshiners and marijuana growers who live hereabouts. As we chatted, other folks stopped in to buy cigarettes, potato chips, or gas. A few inquired whether the Bait Shop was going to stay open beyond Thanksgiving this year. I sat on a plastic lawn chair at the end of the counter, feeling like I should be challenging regulars to checkers. Eventually Harnon and I got around to discussing my purpose in visiting the area. He was shocked to learn that I'd been denied a tour of the dam power plant. As a long-time resident, he said he used go there all the time and see the power plant up close. He even called a friend of his who is a "TVA cop" to see if a favor could be granted for me. Naturally the TVA cop said no. He suggested I call Jeff Ring.

Hoping Ring wouldn't get wind of what might seem an attempted end-run around TVA hierarchy, I said "so long" to the group at the bait store. Harnon walked me out so I could meet his beagle, Lucy, who was tied up near his Honda Gold Wing motorcycle. The pair demonstrated how Lucy could jump up onto the bike's seat by herself. After Harnon and I exchanged e-mails and promises to write, I went back to my room with its pleasant balcony. As I looked out at that massive reservoir, a line from Samuel Taylor Coleridge's "Rime of the Ancient Mariner" played through my head: "Water, water, every where, / Nor any drop to drink."

The copious water isn't all in Kentucky Reservoir, or behind other dams on the Tennessee River system. A map of the Tennessee Valley shows the river looking like a blue snake that swallowed a succession of elongated pool balls. The map reminded me of how many rivers in this part of the country share the name with the state of their origin. Tennessee, Ohio, Kentucky, Illinois, Missouri, Arkansas, Alabama, Mississippi—these names signify both rivers and states. I wondered which came first. So I put that question to John Logan Allen, emeritus professor of geography at the University of Wyoming.

He explained that European settlers in the eastern part of what is now the United States learned American Indian names for natural features like rivers, and adopted them. Those names were kept, even as European settlers moved into the interior of the country. "American Indian names for rivers—partially because of the significance of rivers as avenues of travel in the early exploration and settlement of what is now the United States—were often retained as the names of states through which they flowed," he told me. And this naming pattern wasn't confined to rivers in the southeast. State names like Connecticut, Wisconsin, Illinois, and Michigan are also of American Indian origin.

I drove through some of these river systems on my southern trip, through forested regions that benefit from the generous rainfall in this part of the country. All this moisture and vegetation makes excellent habitat for wildlife. I never saw a deer along the road, alive or dead, on this trip. But the roadkill I cataloged included opossum, armadillo, cats, dogs, turtles, and, shockingly, one bald eagle. I could see its mate sitting at the top of a bare tree, looking down into the highway median at the pile of white and brown feathers that had once been its companion. I wondered what would come along to make short work of the eagle carcass, or die trying.

Once through Tennessee I made my way into Mississippi. I crossed a bridge over the Tallahatchie River. That's the bridge immortalized in Bobby Gentry's song "Ode to Billie Joe." If anyone knows what he threw off the bridge into the river before he eventually leaped off it himself, they aren't saying. But perhaps someone in the TVA could speculate. Their reach extends this far: residents of the Tallahatchie Valley receive their electricity from the TVA hydroelectric plant at Sardis Dam, just west of town.

I crossed the Mississippi River at Refuge, Mississippi, and headed into Arkansas. I was on my way to the Texas gas fields and had chosen a route from Texarkana due west, along Highway 82. My route would take me parallel with the Red River, which divides Texas and Oklahoma. It occurred to me that I was leaving the land where Indian names for rivers were held over to become names of states. Neither Texas nor Oklahoma have eponymous rivers, though a seven-mile stretch of the North Canadian River that runs through Oklahoma City was renamed for that state in 2004. Neither, for that matter, do New Mexico, Arizona, Nevada, California, Utah, Oregon, Washington, Idaho, Montana, Wyoming, Nebraska, North Dakota, or South Dakota. An exception is the Kansas River, which is also commonly called the Kaw, for the Indian tribe.

If you look at a map of the western states that do not share a name with a river, you'll see one notable exception: Colorado. The river we think of by that name was once referred to as the Grand. Then in 1921 the river was named the Colorado, at the request of that state's government. The Colorado's headwaters form in the Rockies in northern Colorado. Its course runs approximately 1,450 miles, through six states and two Mexican provinces. It was once popularly known as "the river too muddy to drink and too wet to plow." But these days it faces so many impoundments on its way to the Gulf of California and the Sea of Cortez that the water might as well be traveling in Morse code, dotting and dashing across the desert.

Numerous dams in Colorado and Utah capture water for various reasons. A big reason is hydroelectric power, and that purpose takes on a serious bent as the river reaches the Glen Canyon Dam. The reservoir it creates is Lake Powell, fashioned by flooding the Glen Canyon on the border between Utah and Arizona. The dam was completed in 1963 and houses a 450-megawatt power plant. It has eight penstocks taking water past the dam turbines.

The Colorado passes through Glen Canyon only to be arrested by another impediment: the Hoover Dam on the Arizona-Nevada border. The reservoir it creates is Lake Mead. I spent a few days at a hotel and casino on the lake, just four miles from the dam. My first night there I watched a newsreel documentary about the construction of Hoover Dam, which plays constantly in a little side room off the casino. Above the clamor of slot machines and 1990s rock music pumping through the PA system, a narrator describes the construction project, voiced over footage of the construction.

From my room I could see a southwestern finger of the lake. U.S. 93, which passes by the hotel, was still busy this November with tourists traveling to the dam, then south to Kingman, Arizona, and on to the Grand Canyon. The road winds around, curves over the dam, and keeps a speed limit of fifteen miles per hour in spots for the safety of throngs of pedestrians. The road isn't practical for automobile travelers hoping to reach anywhere in a hurry. That's why a new bypass road is being constructed, spanning the Colorado immediately below the dam.

Like Glen Canyon Dam, the Hoover Dam was created to stabilize the Colorado River so that its waters could support a boom in the population of folks who wanted to live here. Originally called the Boulder Dam, Hoover Dam's construction started in 1931. The river was detoured from its channel in 1932 so that the several thousand workers could blast the Black Canyon into shape and begin pouring concrete. They poured 6.6 million tons of concrete to build a dam that is 726.4 feet high and 1,244 feet in length. Storage of water in the newly created Lake Mead started in 1935 and creation of electricity began in 1936. The work was partly financed by groups in the six Colorado Basin states who signed contracts for the electrical energy that would be produced, according to the newsreel.

Because recent drought conditions have affected how much net power Hoover dam actually creates, I was confronted with contrary information about the dam. I wanted to be sure to have the most current statistics so I contacted Robert Walsh with the Bureau of Reclamation, Lower Colorado Division. He helped me understand the terminology to get a handle on how power is described throughout the industry, not just in the hydropower field. He reminded me that the term "rated capacity" is sometimes also referred to as "nameplate capacity" and refers in general to the maximum amount of electricity, in megawatts, that a generator or generating plant is capable of producing under optimal or specific conditions. "The generating units at a hydropower plant are designed to operate at a range of pressures," he said. "The amount of pressure placed on the turbine is largely dependent on the depth of the water in the reservoir; all things being equal, the greater the depth, the higher the pressure. The generating units operate at maximum capacity when the optimal amount of pressure is placed on the turbines."

So the rated net capacity of Hoover Dam—the amount of electricity it could generate at once if all seventeen commercial generators were operating under optimum conditions—is 2,074 megawatts. But Lake Mead, which supplies the water to the generating units, is currently about 125 feet below its maximum operating level, Walsh said. So there is less pressure pushing on the turbines. Because there is less pressure, the generators are currently operating at a lower capacity. Because each generating unit is operating at a lower capacity, the rated capacity of Hoover Dam today is only about 1,656 megawatts. If the lake comes back up, the rated capacity will too. If the lake continues to fall, so will the rated capacity, Walsh said.

Unlike TVA management, the Bureau of Reclamation seems to put almost as much energy into dam tourism as it does into creating power. I arrived late in the afternoon to the dam, barely in time for the last power plant tour. About thirty of us paid our eleven dollars each for the company and elucidation of a fellow named Bruce. A retired schoolteacher, Bruce had been giving tours of the Hoover Dam to groups like ours for more than eighteen years. He packed our collection of parents, tired youngsters, romancing couples, Japanese tourists, and power nerds into an elevator like pool balls into a rack. Down we went into the bowels of the dam, 530 feet through the rock wall of Black Canyon. We emerged and shuffled along a tunnel dating back to the time of dam construction into a room with an illuminated diagram board. Bruce described how the water flows through the intake columns through penstocks and past the generators to create electricity. Then back down the tunnel and into the elevator we went to the power plant balcony. From there we could see eight of the dam's seventeen generators. While some of the tourists slumped on benches too far back to clearly see the floor, I and a few other enthusiasts crowded the rail to photograph the 650-foot-long wing of the building.

From there, we were led outside to an observation deck to take it all in. We could see the lake behind the dam, together with the Colorado whose water had done its duty and was allowed to become a river again. It is hard not to wonder what happens to any creatures moving through Lake Mead when they unexpectedly hit the wall of the dam. Like a guillotine, the dam cuts off the system of water and silt and sediment that would normally flow downstream. Silt deposited in riparian areas provides wetland habitat for wildlife. Silt left behind the dams stacks up, making the reservoirs shallower, to the point where they could eventually fill in and ruin the dam. I asked one of Bruce's counterparts, who was stationed on the observation deck, whether fish ever get caught up in the dam. "They don't," he replied. "We pull water in from the reservoir at a location way deeper than where the fish are swimming." He added that trash racks on either side of the river stream remove any large debris that might be floating along the reservoir, before it gets caught up in the dam apparatus.

I haven't heard of any efforts to decommission Hoover Dam and consequently drain Lake Mead. However, there is a movement to decommission the highly controversial Glen Canyon Dam and release Lake Powell back into the Colorado River. The Glen Canyon Institute (GCI), formed in 1996, is one of the more high-profile organizations making this argument. It quickly gained the support of David Brower, who served on its board before he died in 2000. Brower served as the first executive director of the Sierra Club and is credited with preventing Dinosaur National Park in Utah from becoming a lake. Those who fought against flooding Dinosaur at the time compromised and decided not to fight Bureau of Reclamation's dam at Glen Canyon. Later Brower would call that dam "America's most regretted environmental mistake." That puts the controversy about this dam near the level generated by the Tellico Dam on the Little Tennessee River.

The GCI continues to advocate for the dam to be decommissioned, not removed, which they say would be too costly. Instead, they call for the waters of Lake Powell, which they refer to as the "dead pool," to be lowered and to flow through a newly created channel around the dam. They say the "bathtub ring" created by the stain of Lake Powell's waters would diminish and the Colorado River's health could be restored by being allowed to flow freely all the way to Lake Mead. The GCI believes the geography of Lake Mead, downstream, is much better suited to handle the problem of sediment buildup that could one day choke Lake Powell. They point out that the waters of Lake Powell serve one small community and a coal-fired power plant on the Navajo Indian Reservation. They believe those needs could be served as well by Lake Mead.

As to electricity, the GCI says the power created at the dam could be replaced with "simple conservation measures and expanded cooperation with power distributors." They voice an opinion about hydroelectric power that recalls the view of the Save Our Wild Salmon group, which hopes to remove run-of-river dams on the Snake River in Washington. "Glen Canyon Dam doesn't generate 'clean' power," states the GCI. "While there is no air pollution from the dam, the 186-mile-long reservoir (which serves as the 'fuel' for power generation) has destroyed one of the most incredible regions in the world. Add to that the batteries, feces, and gasoline dumped into the reservoir each year by the myriad speedboats and houseboats. In a hundred years, when the reservoir is filled with toxins and sediment and well beyond any hope of restoration, which will have been the "dirtier" power source?"

In his essay "Living Dry," another high-profile environmentalist, Wallace Stegner, writes: "Adaptation is the covenant that all successful organisms sign with the dry country." That phrase is affixed to the wall of the Alan Bible Visitor Center at Lake Mead. It is apparently intended to help visitors understand the creatures that inhabit this country of extreme heat and very little water. The visitor center gallery displays interactive images of Mojave residents such as the desert tortoise, the big horn sheep, and the raven. One can press a button and hear various recorded caws, squeaks, rattles, and chirps of the animals. At certain times all of the sounds fill the air at once, as though the animals have converged at a town hall meeting.

In the center of the room is a relief model of the region around the Lake Mead Recreation Area, including Lake Mead and Lake Mojave, formed downstream by Davis Dam. Visitors can push a button and illuminate with tiny light bulbs what they want to learn more about. Possibilities include where emigrants ferried across the Colorado, where small towns once stood that are now under the lake, and where railroads were built to support dam construction. The lights are brightest when one presses a button for Las Vegas. Lights spread across the map's valley floor illuminate how access to water and electricity open the door for population expansion. The population of Las Vegas is now at about 1.9 million. Hoover Dam generates enough power for about 1.3 million people, with its customers in Vegas and around the region.

If adaptation is essential for living in a dry country, as Stegner says, then humans have found a high form of adaptation, through engineering. Stegner's choice of the word "covenant," which means a solemn agreement binding all parties, makes me realize we all have something at stake, something to gain, and something to lose, by controlling nature in this way and in this place. The sound of human voices might be the next element added to the interactive display of creatures at the visitor center. It is by far the loudest.

8

Don't Let the Sun Go Down . . . without Capturing Its Energy

A Nevada Solar-Thermal Power Plant

NOVEMBER 1

It was 39 degrees in Laramie today, several degrees warmer than the average for this time of year. With a low of 29, also higher than average, and only a light breeze, it felt like spring in the high country. I was headed home after a trip to some of the vowel states of the Midwest: Iowa, Indiana, Illinois. I'd paused for a few days in the double-consonant state of South Dakota and was glad I did. In my absence Laramie received its traditional Halloween storm, a few days early. About a foot of snow dumped on our region, closing I-80 for several days, not just across southern Wyoming but east into Nebraska as far as North Platte. While I holed up in Vermillion, my friends to the west dug themselves out, costumed their young children, and doggedly trick-or-treated in parkas and snow boots.

By the first of November the snows retreated and the birds poked their heads back out. I made the trip back to Laramie on the Sunday that marked the end of Daylight Saving Time. The sun rose at 6:34 a.m., which I was not there to see. But I was heading down the 8,471-foot mountain pass into the Laramie Valley as the sun was setting. That was at 4:57 p.m. I was right on time to see a postcard full moon low over the

Sherman Mountains as I switchbacked the twelve hundred feet into town.

The next few days were indeed a return to spring. The weather flip-flopped this year, with October weather suited for a Christmas tree. Football Saturdays at Wyoming Cowboys' games found us in down ski pants, with foot and hand warmers in our boots and mittens. The cottonwood and crabapple tree leaves we were ready to rake and bag were abruptly buried under several inches of snow. But after the Halloween storm the weather moderated. We broke out the outdoor grill. The usual winter visitors, chickadees and nuthatches, were joined by unseasonal passersby such as goldfinch, pine siskins, and grackles. It felt like we were headed to the soft side of spring, yet I knew better. I wondered what was going through the mind of the lone robin that had been hanging around the last few days. He feasted on ripened cotoneaster berries. He drank from the birdbath, the horse trough heater we employ to melt the ice not yet needed.

With the shorter days and earlier nights the little plastic solar lights we have stationed along the fence line and around the xeriscape garden were doing longer duty. As a result, the batteries that keep them able to work blinked out, one by one. Even in the shortened days of winter, it seems the sun is almost always shining in Wyoming. That's one reason people from vowel states and elsewhere come here in the winter, to ski in the mountains and get a fix for seasonal affective disorder. There's enough sunlight for rooftop photovoltaic solar panels to be a good option in some areas of the state. Yet Wyoming does not have enough sunshine to make it a viable location for the type of concentrated solar energy that goes out on the power grid. The Energy Information Administration describes concentrated solar energy potential as the capacity to produce a minimum of 6.0 kilowatt-hours per meter squared, per day. An EIA solar resource map shows in yellow only a few spots in southwestern Wyoming that achieve that magic formula.

Colorado has a bit more yellow on the map, about the same as Texas and Utah. New Mexico has more still, but the solar capacity of Southern California, Arizona, and Nevada colors huge swatches of these states yellow.

Nevada is the number one state in the nation in solar energy generation per capita, and I arranged to pay a visit to a new solar-thermal facility there that generates commercial-scale power. But first I had to get the terminology straight. Solar-thermal power uses direct heat from the sun to boil fluid for steam. The steam spins a turbine, which turns a generator, and electricity results. There are several types of solar-thermal plants. They all concentrate the sun, but with different technologies. These are the linear, the dish/engine, and the power tower concentrators. The other main variety of solar power, photovoltaic, converts sunlight directly into energy. Rooftop solar panels use the photovoltaic system. Panels on the roof; power to your blender. Both of these technologies are being used to create commercial-scale, grid-compatible power in the United States, although photovoltaic does so on a smaller scale.

The plant I arranged to visit, Nevada Solar One, is a concentrated solar-thermal power plant of the linear concentrator sort. The facility is located approximately thirty-two miles southeast of Las Vegas in the Eldorado Valley. I had never been to Las Vegas. I love the idea of Frank Sinatra, Dean Martin, and the rest of the Rat Pack setting up a Hollywood outpost in the desert. But I am not a fan of crowds, bright lights, heavy traffic, overproduced entertainment, and above all, the incessant boop boop boop of electronic slot machines. But to get to Nevada Solar One by plane was to go first to Vegas. I flew from Laramie to Denver, boarded lucky Frontier Flight 777, and headed southwest.

From the air over southern Utah I could see the canyon country around Lake Powell, formed from impounding the Colorado River. A few minutes later I could see Lake Mead, which also impounds the Colorado. I had long felt disturbed that the river's bloating through these parts had drowned so many canyons, washing away important habitat and artifacts of lost cultures. But when the plane entered the airspace over Las Vegas, I caught sight of another sort of bloating. Pavement, buildings, swimming pools, and rooftops, mostly without photovoltaic systems, filled my view. If it weren't for the mountains, the Vegas area could eat up even more desert in its rush to sprawl.

When the economy slowed in 2008, one of the fastest-booming American cities trimmed its growth rate down to that of cactus, temporarily at least. Still, Las Vegas is the biggest city in Nevada, with a metropolitan area of 1.9 million people, as of 2010. That makes it the twenty-eighth largest city in the nation. According to the U.S. Geological Survey, Las Vegas is the brightest city on earth, as viewed from space. Powering the lights on the Strip alone requires 4.5 million kilowatt-hours a day. I didn't see how a solar-thermal power plant like Nevada Solar One could make any worse use of desert landscape than a megahotel on the Las Vegas strip.

I took pains to find lodging outside of Vegas and located the Hacienda Casino, southeast of Vegas, between Boulder City and the Hoover Dam. Boulder City was built by the Bureau of Reclamation in 1932 to quarter workers constructing what was then called the Boulder Dam. After the job was completed, workers and others realized that the beautiful setting and warm dry climate would suit a permanent town. Boulder City flourished and Reclamation gave up control of it in 1958. The town was incorporated in 1960. It continues to uphold the portion of the original town charter declaring it gaming-free, but tossed out the ban on alcohol.

The Hacienda Hotel and Casino is outside of the city limits. It allows both gaming and drinking, not to mention smoking. I had resolved to spend twenty dollars on a slot machine, thinking that forcing myself into a new experience would be character building and might even yield a down payment on something I didn't yet know I needed. I furtively threw a few quarters into a machine and discovered that winning didn't mean receiving clattering coins that I could pocket. It meant getting more chances to lose. After about two minutes of feeling extremely stupid, I resolved to spend the remaining nineteen dollars on a Hoover Dam T-shirt, or anything besides these slot machines. The house was safe, from me at least, and I left the casino for the quiet of my room, well above the casino's fray and in the nonsmoking section.

I had arranged to meet Michele Rihlmann-Burke, Acciona Solar Power's marketing and communications manager, early the next morning. I was very happy to have worked out this tour. It was one of the most difficult to arrange of all my travels. I had to explain to folks higher up in the company why I wanted to visit and explain my stance on renewable energy, particularly solar. I gave them the link to my blog, to assure them I wasn't going to (a) waste their time; (b) write poison pen missives about solar power; or (c) try to blow the place up. I identified these three potential concerns only through sheer speculation. They never did tell me any specific reasons for their hesitation. To Acciona Solar's credit, I had first contacted other solar power businesses, including Florida Power and Light, which owns the facility at Kramer Junction, California. They all told me in no uncertain terms, "We don't give tours." Also to Acciona Solar's credit, my host agreed to drive more than an hour each way from the corporate office in Vegas to meet me and spent half a day in my company—a significant investment in time.

When giving me directions, Rihlmann-Burke told me that because there was nothing else out there in the desert along Highway 95, Nevada Solar One would be easy to spot. I left my room at the Hacienda at about 7:00 a.m. I was spurred into early departure by a combination of my internal clock being set on Mountain Time and my anxiety over not knowing where I was going. Images of getting lost at Kentucky Lake Dam swirled around in my head.

In spite of my host's claim that nothing much was out there, I noticed many things in the desert while searching for Nevada Solar One. Chief among them were the high power transmission towers taking the electrical load from Hoover Dam into the grid. The towers were so large I imagined the solar plant to be to that scale. I've lived the last twenty years of my life in places where the landscape is dominated by mountains, forests, or livestock, things one easily notices from the road. But as I scanned across the Mojave Desert through my windshield on this overcast November morning, I started to wonder where the creosote bushes and sand ended and where the power plant began. I expected to see glittering mirrors on tall frames but instead saw only a low area off to the west that made the desert appear whitewashed. It didn't look like any power plant I'd seen yet.

Fortunately, a well-marked turnoff to Eldorado Valley Drive signaled the entrance. The road serves the twenty-eight or so workers at the solar plant together with activity at the Eldorado Energy natural gas-fired power plant just beyond the solar field. After I located the solar plant I took a look at my watch and map. I still had about an hour to kill. A road sign on two-lane Highway 95 indicated the town of Searchlight about twenty-five miles to the south. I decided to kill time as a windshield tourist: drive down there, get a quick feel for the place, and drive back. I pointed my rental car south through the valley, the Eldorado and River Mountains framing my view.

Creosote bushes and Joshua trees were the primary flora I could see; fauna steered clear of the road, at least on this morning. Sandy alkaline soils contrasted with the dark volcanic mountains that seemed plopped on top of the earth's surface like chocolate pudding, hardened in the sun. I crested a low hill and saw the Mojave National Preserve off to the southwest. Suddenly the hour I'd wondered how to kill seemed frustratingly brief, as I had no time to really explore. Worse, my stop in Searchlight was not much more than a drive-through. This former gold-mining town has about five hundred residents, most famously U.S. senator Harry Reid. Searchlight has few public facilities, but naturally there is a combination convenience store slash casino, where I stopped. I picked my way past the slot machines to find something I could justify buying. I dished out 99 cents on a pack of roasted peanuts just so I wouldn't look as uncomfortably stupid as I felt.

After I made my big purchase in Searchlight I headed back toward Nevada Solar One. Just south of the Eldorado Valley turn I saw a sign for an area of critical environmental concern. That means the Bureau of Land Management oversees this federal land with particular attention to, in this case, the habitat of the endangered desert tortoise. One of the most iconic species of the area, the desert tortoise digs its burrows underground, where it spends the majority of its time out of the hot desert sun. As development has displaced its habitat, efforts have been made to relocate the tortoise. According to the group Defenders of Wildlife, this effort usually fails. Turns out the desert tortoise doesn't want just any old burrowed tunnel. It wants its own and will die trying to get back to it.

More than 85 percent of Nevada is BLM or other types of federally managed land, far more than any other state. That's because when the western territories became states, Nevada decided to sell most of its land back to the federal government and take the cash. Compare that amount to Alaska, the state that comes in second in public ownership, where the federal government manages just over 60 percent of the land.

Public land is for use by all of us, but that does not mean there are no rules. For everything one might propose to do on public land, one must follow specific regulations. There might be rules about where to camp, or where to drive on or off road, for example. If one wishes to develop a solar plant or wind farm, a variety of federal and state regulations must be followed during the permitting process, including environmental impact studies and public hearings to allow locals to give input on that planned development.

In 1990 Boulder City, the state of Nevada, and the federal government began a process to transfer 107,412 acres of Eldorado Valley to Boulder City, completing the deal in 1995. Known as the "transfer area," the land was to be used as open space for tortoise preserves, recreation, and for developing solar power plants. Boulder City designated six thousand acres of its newly possessed territory as a solar enterprise one. Their plan was to lease this land to a variety of companies developing solar energy. Some area residents were surprised when one of the first energy endeavors built there was the Eldorado Energy gas plant. Operated by Sempra Generation, it also has a 10-megawatt solar project on its grounds, so technically it deserves to be there. It shares a power substation with Nevada Solar One.

I wanted to know why Boulder City thought it important to include a solar enterprise zone in their transfer land, so I contacted Rose Ann Rabiola Miele, a jack-of-all-trades and the town's public information officer. Her voice over the telephone pegged her as a nonnative of Nevada. Indeed, she said she had lived in Chicago for forty-five years "in Richard Daley's neighborhood" and relocated to Boulder City about fifteen years ago. Not only is she in charge of public information, she runs the town television station and hosts a local program. She told me she works from seven each morning to at least six each night and her cell phone is always on. At sixty-two, Miele said she looks forward to retirement in a few years, not surprisingly, so she can "do some other things."

I told Miele I had watched a National Geographic documentary about the building of Nevada Solar One. The narrator enthused about how the power from that plant would be used to light Vegas, especially the Vegas Strip. Like it or not, since power is demanded, it might as well come from a renewable source, I thought. Yet I wondered if there might be hard feelings that power from their land was going to their gluttonous neighbor. As it turns out, Boulder City isn't that concerned about those other destinations for the power bought from Acciona Solar Power. That's because the city receives millions of dollars each year from leasing the land. The city projected that in 2011, it would receive $2.5 million in leases from projects on the solar enterprise zone, alone. Also, water is required by the solar plant to boil into steam to create electricity, and to keep the mirrors clean. That water comes from the municipal supply at Boulder City and isn't given away free, either. In addition to water for the power plants, the city sells water to a golf course in town that is owned by Harrah's, the big Las Vegas entertainment outfit. Miele explained that the town was supportive of the Las Vegas Strip and making sure it didn't go dark, so to speak.

I ruminated on this information on my way to the plant after my peanut run to Searchlight. I approached the plant from the south listening to local talk radio and admittedly zoning out a bit on the lightly traveled highway. The pudding-colored mountains had turned from chocolate to butterscotch in the hazy morning sun. Nothing looks the same in the desert when seen from the other direction. Awash with a creeping panic that I'd missed Eldorado Valley Road, I was relieved to finally see a sign indicating I was in Boulder City, at least the exurban, solar enterprise zone part. I found my turn, headed west, and drove a mile and a half to the plant gate. Picking up the phone in a box, I announced myself like a seasoned power tourist and drove to the operations and administration building to meet my host.

Michele Rihlmann-Burke had recently moved to the Vegas area, leaving behind a job in California with an interactive company that developed websites for sports teams and other clients. The male-dominated power game was a new one to her, and she seemed up for the challenge of showing the plant employees we met that she had grasped and could clearly communicate the nuances and complexity of this form of renewable energy.

Our first stop on the tour was to get hard hats and eye protection. As we rummaged through a large box for equipment, Rihlmann-Burke explained how the power plant came to be built. Because the land is owned by Boulder City, Acciona faced fewer regulatory hurdles than they would have if building on BLM-managed land. As it is, the plant represents a $266-million investment by the company. That's twice what it would cost to build a new coal-fired power plant, but then again, there are no fuel costs to pay. The power the plant generates is sold to Nevada Power Company and Sierra Pacific Resources under long-term power purchase agreements. A PPA is a financing device between private developers and public-sector partners. Developers get tax credits; power customers get energy prices at a set rate.

Rihlmann-Burke took me into the control room where we met the control room operator, Aaron Boucher, and a technician, Matt Haines. These two sat among the various computer monitors that showed mirror positions and temperatures of heating fluid moving through the tubes. Boucher explained how concentrated solar works at Nevada Solar One. Though designs vary, the system uses mirrors or sometimes lenses to capture sunlight that shines across a large area and concentrates it onto a small area, in this case, tubes of an oily transfer fluid. That fluid is encased inside a pipe called a solar receiver. While in the solar receiver, the fluid heats to several hundred degrees. The fluid then passes through a gas heat exchanger inside the power block. There the fluid becomes even hotter and turns into steam. The steam drives a conventional turbine connected to a generator that produces electricity.

When Rihlmann-Burke and I left the control room, the fellows were still gazing dolefully out the control room window at the mirrors, dull in the overcast sky. They could control the direction of the mirrors to track the daily progress of the sun, but they couldn't do much about the clouds that hamper power production. The plant has a maximum capacity of 75 megawatts. Even on cloudy days like this one, it can still operate. That is because the heated fluid that has already entered the plant can still do its job, for a time. Nighttime is another matter, especially during the short days of winter. The thermal storage technology that would allow solar power plants to store electricity created on sunny days is still under development. Material called molten salt is being tested because it can retain heat, Rihlmann-Burke told me. Theoretically it could retain heat made by the solar plant during the sunny day and release it at night, allowing the plant to continue making steam, she said.

We stepped out onto a balcony and took in the panoramic view of this section of the Mojave. As I had been reminded out on the highway, life in the desert takes on subtle form and it can take a careful eye to observe it. But here, the sandy soil really is lifeless, with thousands of mirrors garrisoned between it and the sun. Rihlmann-Burke pointed out a patch of desert sand beneath and around the mirrors that appeared to be wet. There was a shimmering quality to the ground because it is regularly sprayed with an oily substance. Not only does the substance keep down dust and dirt, it also keeps anything from growing or otherwise living under the mirrors. Dust and dirt are major inhibitors of the reflective power of these mirrors. What dirt does still drift up to the mirrors in spite of the treated ground sends the crew outdoors for a regular mirror washing.

"About 5 percent of the plant's water use is for washing the mirrors," Rihlmann-Burke explained. "Also, more washing occurs in the summer due to winds and dust that can accumulate if regular cleaning isn't done."

And while it doesn't often snow in the Mojave Desert, when it does, it benefits the plant. "We've found that snow is a great way to clean the mirrors, so believe it or not snow is good for concentrating solar," Rihlmann-Burke said.

We left the balcony and wandered among the temporarily underachieving mirrors. Sandy soil crunched underneath our boots and the solar mirrors creaked like old bed springs: otherwise there was no sound, not even from distant Highway 95. We got a close look at the mirrors, arranged on aluminum frames to form a trough. Nevada Solar One uses 760 parabolic concentrators and more than 182,000 mirrors that concentrate the sun's rays onto 18,240 solar receivers. The mirrors and troughs take up about three hundred of the four hundred acres of this facility. They are built using "white" sand, so called because the material contains no iron. Most mirrors have a greenish tint because iron is present, Rihlmann-Burke said. Not only are these mirrors white, they are extremely thin. They are manufactured by Flaberg, a German company that specializes in automotive glass. Acciona's order for solar mirrors prompted Flaberg to retool one of its glass factories to meet that demand. Acciona hopes that if the solar industry can demand enough of this sort of glass, the glass manufacturing industry in the United States will dedicate plants to the purpose, bringing down costs.

Nevada Solar One is the third-largest solar concentrating plant in the world, Rihlmann-Burke told me. The facility was built over the course of sixteen months, with the help of eight hundred construction workers. It went online in June 2007 and generates electricity for the equivalent of fourteen thousand households in Nevada. It was the first commercial-scale power plant to be built in this country in over twenty years. The well-known solar fields around Kramer Junction, California, deploy their mirrors in similar parabolic troughs. But the efficiency of the solar receivers and the engineering of the mirrors have improved greatly over the years since that facility was built.

Flanked by long rows of mirrors in frames, we strolled under cloudy skies about a quarter mile to the power block. Although Acciona Solar had allowed me to visit the facility, management would not allow me to enter the power plant itself. I had no reason to doubt Rihlmann-Burke's word that the plant is loud but very clean compared to a fossil-fuel plant. From our position outside, she pointed out the exterior pipes where the transfer fluid enters the plant. We could see the process Boucher had explained in the control room: the fluid flowing through a gas-fired heat exchange and then turning to steam. Also from outside we could see the cooling system of the plant, composed of a trio of fan-driven cooling towers that allow the steam to return to its liquid water state and be used again in the process.

"Nevada Solar One uses wet cooling," Rihlmann-Burke explained. "Dry cooling is a great solution where water is scarce." As I wondered what could be a dryer location than the Mojave, she continued. "At the time of construction, wet cooling was the more economical choice—it's cheaper to install than dry cooling. Plus, water is available at the Nevada Solar One site, making wet cooling an equally viable solution." The water she referred to is the supply from Boulder City.

Water use is probably not the first drawback to solar energy that people would think of. But it is one of the many concerns that environmentalists raise when arguing against commercial-scale solar plants in the desert. In the case of Nevada Solar One, recreationalists and environmentalists had various bones to pick. But ultimately, Boulder City's goal of incorporating solar development into its land fit with its stated goal of supporting renewable energy.

Indeed, according to Acciona, the "strategic implementation" of concentrating solar plants in the United States could "augment and eventually replace" fossil-fired power plants like the gas plant out the back door of the operations and administration building at Nevada Solar One. That would lead to a zero-carbon grid within just a few decades. The company says the United States offers several hundred gigawatts of concentrating solar power potential; however, concentrating solar power accounted for only 1 percent of all energy consumed in the United States in 2008. One concern about commercial-scale solar power plants is the amount of space they take up. But solar advocates like Acciona argue that this is a good use of space and that a concentrated solar-thermal energy project deployed over an area less than one hundred square miles in the Southwest could produce enough electric power for the entire country.

As Brian Hayes, author of *Infrastructure: A Field Guide to the Industrial Landscape*, notes: "It's sometimes said that to run the country on solar power we'd have to pave the whole landscape with collectors. It's not nearly that bad. According to one estimate, photovoltaic plants that could meet the electricity needs of the United States would occupy a little less than 12,000 square miles. That's a lot of land, but it's only about one-third of one percent of the total land area of the nation. So there's no need to pave over the whole country, just the state of Maryland."

Maryland is safe because it doesn't have much commercial-scale solar potential, but states like California and Nevada are ripe for development. As long as desert areas such as Nevada's are valued more for what they can contribute to the greater good than for what they are, development is likely in many areas. Part of Nevada was already sacrificed to the U.S. Department of Energy, which tested nuclear weapons underground at the Nevada Test Site. That location is seventy-five miles northeast of Las Vegas. Radioactivity from nuclear materials entered the ground and into aquifers. When testing ended in 1992, the Department of Energy estimated that more than 300 million curies of radiation had been left behind, making the site one of the most radioactively contaminated places in the country. For comparison, the accident at Three Mile Island released 13 million curies into the atmosphere in 1979. And the Japanese government estimates about 60 million curies of radiation were released into the atmosphere following the

2011 earthquake and tsunami affecting the Fukushima Daiichi nuclear plant.

These are the sorts of complexities people would do well to understand before betting on forms of energy as if they were sports teams. Solar and other renewable energy advocates, just like backers of traditional energies, have argued for their preferred forms of power. Hardly ever do these advocacy messages mention the weaknesses of their "team." Their advocacy messages appear to have paid off, however, in terms of public opinion.

In 2009, Gallup's annual energy poll showed that more than three-quarters of Americans surveyed would like government funding for renewable energy to increase. In another survey, the Solar Energy Industry Association engaged the independent polling firm Kelton Research to consider the energy issue, specifically solar energy. The survey conducted in October 2009 showed that more than 90 percent of Americans would support solar developments and funding over any other renewable source. When asked what one energy source respondents would financially support if they were president, 43 percent said they would choose solar. Wind pulled in 17 percent, natural gas garnered 12 percent, and nuclear came in at 10 percent. However, many people said they would like more information. While about 12 percent of those polled said they are extremely informed about solar power in general, 74 percent said they wished they knew more about solar options for their home or business.

Clearly, most people feel positively about solar power. Maybe that's because the fuel is free. Or maybe humans crave sunshine. Or maybe it's because the news about the drawbacks of commercial-scale solar power hasn't reached many listeners. It has been said that people care deeply about environmental issues affecting landscapes that are green, such as rain forests. But if a place is brown, like a sage-covered basin or desert, its ability to convey pathos is diminished by several shades.

Out in the desert, other types of commercial-scale concentrated solar plants under development use technology that is a bit different than that at Nevada Solar One. I'd learned what they were in order to prepare for my trip, but only after seeing Nevada Solar One was I really able to picture the differences in technology. Some use parabolic dishes, rather than troughs, to concentrate energy. Others employ a central receiver to concentrate energy. That type is called a power tower, because it is a tall monolith ringed by mirrors on frames, known as heliostats. The mirrors concentrate the sun to a spot at the top of the tower, causing fluid in the tower to attain a high temperature. Then that fluid flows into the power plant to create steam and turn a turbine to create electricity.

An example of this technology is the Ivanpah Solar Electric Generating Station, scheduled to be built in California's Mojave Desert. It has three concentrating solar-thermal power plants. Each plant will use thousands of seven-by-ten-foot sun-tracking heliostats. Development of solar plants is one objective in reaching California's goal to boost its use of renewable energy to a mandated standard of 33 percent by 2020.

However, not everyone in California or the region supports commercial-scale solar development, or wind, or fossil-fired plants, or nuclear plants, or biomass fuels, or hydroelectric power. In this case, critics note that construction of the plant at Ivanpah would require grading more than six square miles clean of vegetation, leveling one-hundred-year-old cacti and creosote along with rare indigenous plants. Some people view this as unacceptable environmental harm; others see it as sacrificing a bit of a vast landscape to prevent worse destruction in the form of greenhouse gas emissions if a coal-fired plant were to be built, instead.

Like other types of renewable energy, solar power came into vogue in the United States in the 1970s with the wave of interest in anything but foreign oil. Thank the oil embargo of 1973 for the spike in houses with black photovoltaic collectors planted on rooftops, soaking up the sun. As tensions with Middle Eastern countries eased for a time and oil became plentiful and affordable, those panels became black albatrosses perched on house roofs. On a few occasion over the years I have toured homes for sale with solar collectors on the roof. I have asked the real estate agent about solar power in the home. In each case, there was shoulder shrugging accompanied by mumbling that no one even knew how the thing worked. It was as if we were archaeologists visiting an ancient culture, seeking to understand its ways by sifting through potsherds and bone fragments.

The technology isn't quite that old, according to the National Renewable Energy Laboratory. In 1839, French scientist Edmund Becquerel discovered that certain materials would give off a spark of electricity when struck with sunlight. Then in 1954, Bell Telephone scientists discovered that the process of converting light (photons) to electricity (voltage) could have some promise. Thus the term photovoltaic. Scientists discovered silicon, which is an element of sand that creates an electric charge when exposed to sunlight. That's when solar cells started to power space satellites, together with watches, calculators, and other small items.

Photovoltaic solar energy has become a popular option for not just individuals but also for communities wanting to produce their own locally grown power. This trend has been supported by the U.S. Department of Energy's Solar America Communities program. In 2007 and 2008, DOE selected twenty-five major U.S. cities that are working to accelerate the adoption of solar energy technologies in their communities. Solar projects in these communities will be eligible for DOE financial assistance.

The DOE says that their Solar America Communities program is "committed to developing a sustainable solar infrastructure that removes market barriers and encourages the adoption of solar energy by residents and businesses in local communities.

The objective is to develop comprehensive approaches that lay the foundation for a viable solar market and provide a model for communities throughout the United States." To be sure, the DOE also has initiatives around other sorts of fuels, "making the most" of traditional fossil fuels while noting that "advances in wind, hydro, and geothermal energy allow us to take advantage of clean, abundant energy."

Whether photovoltaic or thermal, solar power needs room. A commercial-scale concentrating solar power plant, such as Nevada Solar One, needs approximately 8.5 acres per megawatt. That's a lot of land; in fact, it is seventeen times as much land as a nuclear plant needs to generate the same amount of electricity. The Alliance for Responsible Energy Policy, a California organization, is one of many that argue against concentrating solar power's intensive use of land and for distributed generation. They advocate a system of microgrids in which people can easily and affordably install their own wind turbines and photovoltaic roof systems to create electricity for their own power use. Any excess electricity would be sent onto the power grid, via transmission lines that would still connect houses to one another and on to power substations. Depending on a household's use, an electric bill might arrive one month, but the next month might generate a check.

The microgrid can be on a larger scale than just individual homes. At the California Correctional Institution in Tehachapi, California, a photovoltaic system covering two acres provides solar heating. A solar collector field heats a pressurized water district–heating loop to about 240 degrees, providing thermal energy for space heating and domestic hot water at a complex serving six thousand inmates and staff. During the summer months, the system provides essentially all of the thermal needs of the prison.

Popular in part due to efficient use of space, distributed generation would mean photovoltaic systems on land already covered by structures such as individual homes and parking garages. Distributed generation could also extend to other sources of power, such as individual wind turbines powering batteries to run household appliances and lights. In these ways, electrical power could be generated at or near the location where it was needed, rather than pumped out into the grid where it loses efficiency the farther it travels from its source.

Throughout the national debate on energy, one might expect the most heated exchanges to be between the pro–fossil-fuel group and the all-renewables-tomorrow group. But the battle of green versus green is every bit as impassioned. Solar supporters, for example, get shut down by wind supporters because of the unfortunate fact of nightfall. Wind supporters quietly overlook the fact that many people don't want to see wind towers poking up every few hundred feet across ridges, coastal areas, and other viewsheds. And so on. The general theme of the discourse can be summed up as "Sure, we don't want greenhouse gases, but don't mess up my desert, hilltop, view, river, canyon, or backyard."

With the competing voices of power companies, government groups, environmental activists, and consumers, it is easy to lose sight of the fact that it isn't just our homes sucking down power. Even though industry creates tangible products, little mention is made in the public discourse about the industrial process that goes into what we use every day. More than 20 percent of our nation's energy consumption goes to manufacture food, clothing, paper, electronics, automobiles, transportation fuel, railroad cars, airplanes, battleships, and so on up the line, the EIA says.

A whirling windmill in the lawn or some thin photovoltaic cells on a roof can power a home, but it is intermittent power. Heavy industry requires baseload power such as coal or nuclear. It would be a stretch to think that intermittent power sources such as wind and solar would have the heft to run our nation's industry without help from traditional power sources any time soon. Traditional sources might not do well in a distributed generation, or microgrid, model. For example, we won't be having backyard nuclear plants any time soon. But as contributors to all the sorts of electricity pumping along electric lines that crisscross the country, renewables can supplement other forms of power.

When it comes to technological innovation, many facts that are true today are on the verge of not being true tomorrow. For instance, the Solar Energy Industries Association's Jared Blanton noted that industry currently depends on traditional baseload power. "But, solar-thermal technologies are beginning to move into that space," he told me. "Probably the most interesting example of this is this Steinway piano factory in New York. Steinway requires process steam and is now using parabolic trough solar collectors to create it." And that's in New York, not a place that ranks high on anyone's map of great solar resources.

The EIA anticipates innovation when it describes solar's residential and commercial future. "In the future, it is possible that utility-scale photovoltaic plants will compete with wholesale electricity generation, provided that further technological advances are achieved." Solar-thermal power plants are intended to compete with other types of power that are called upon at peak times, such as natural gas–fired plants.

In the 1970s and 1980s, cost restricted the competitiveness of solar power. Today, homeowners have assistance available to them if they want to install solar panels on their rooftops or a wind turbine on their rural lawn. The Solar Energy Industries Association provides links on its website to information about renewable energy, especially solar. So does Energystar.gov. Nevada alone has a few hundred organizations devoted to the manufacture, development, or promotion of solar energy. Almost all states offer some form of grant money or tax credits for installation of solar or wind systems.

And why wouldn't they? We are at a time when people are becoming convinced that the earth is warming and greenhouses gases from burning fossil fuels are part of the cause. But it is the sun, as the Union of Concerned Scientists reminds us, which supports all life on earth and is responsible for the energy we use. "The sun makes plants grow, which can be burned as 'biomass' fuel or, if left to rot in swamps and compressed underground for millions of years, in the form of coal and oil. Heat from the sun causes temperature differences between areas, producing wind that can power turbines. Water evaporates because of the sun, falls on high elevations, and rushes down to the sea, spinning hydroelectric turbines as it passes."

They point out that all of the energy stored in the earth's reserves of coal, oil, and natural gas is matched by the energy from just twenty days of sunshine.

Wow.

I think I'm underutilizing the power of the sun.

I considered that possibility as I flew out from sunny, breezy Vegas, over the Utah canyon country and the snow-covered Rocky Mountains to Denver. By the time I'd switched to a commuter plane and landed in Laramie I'd gained seven thousand feet in altitude and thirty miles per hour in wind speed, and lost 40 degrees of temperature. Once at home I stepped out on the backyard deck and watched the little plastic bluebird that decorates my fence spinning its wings in the wind, a blue blur. I could hear the voice of that practical daughter of Chicago, Rose Ann Rabiola Miele, prodding me, "Look at your electric bill. Most people have no idea how much they are using so they don't know whether conserving pays off."

She was right. Other than looking at the amount owed, I barely glanced at the details. I rummaged for a bill and discovered that in that chilly month of October, I'd used 459 kilowatt-hours. Considering that both my husband and I work at home using computers all day, that isn't bad, but I wondered if it could be better with distributed generation. Alas, information I've seen suggests that installing rooftop solar panels would not be economically beneficial for users at our level.

I continue to use the sun indirectly, at least. I put my house-plants in the window and open the blinds to the morning light. I greet the sun to walk my dog twice a day no matter what the weather, which is usually sunny even when chilly. I get outside and ski, or hike, or soak in the sun reading a book on the backyard deck depending on the season. My problem isn't getting too little sun, but getting too much of it. So I remember to put on sunscreen each morning and remind my husband to put a hat on over his thinning hair. I rejoice that I live in a place where if one gets seasonal affective disorder from insufficient sunlight, one might feel better just through the steps I'm already taking. No matter what source of electricity flows through the utility line attaching my house to the substation then on to the transmission lines beyond, I remind myself it's the sun that makes the world go around.

9

Harnessing the Moon

A Maine Tidal Power Project

DECEMBER 1

The high of 37 in Laramie on this Tuesday, and the low of 11, was a gift for this time of year. Except we knew what was coming. Our low on Tuesday night would be Wednesday's high, and the low Wednesday night would crater to minus 16 degrees. We'd have time to recover from the chill by reaching a high Thursday morning of 4 degrees. But of course the low that night was minus 19. And so on and so forth, all adding, or should I say subtracting, to one of the coldest Decembers on record in Laramie.

I was watching the weather forecast all fall, not just for conditions in Laramie, but in Eastport, Maine. I planned to visit the site where the Ocean Renewable Power Company was testing a new way to coax power from ocean energy. ORPC is developing what is known as hydrokinetic energy—not by damming currents—but by placing water-driven turbines into oceans and rivers. The beta version of their turbine generator unit (TGU) they were testing in Cobscook Bay at Eastport was the largest ocean energy device deployed in U.S. waters thus far. Testing was progressing by fits and starts, as is the way with most new technology. I'd hoped to visit in September, when weather in coastal Maine is typically fine. But ORPC's testing schedule, and my teaching schedule, conflicted. Before I knew it I was looking at making a trip in December.

That idea brought to mind a television commercial from long ago for frozen fish sticks. It featured yellow-slicker-clad fishermen clinging to the rails of a rocking fishing boat on fifteen-foot seas in what the voiceover announcer described as the "Icy North Atlantic." Maybe that's not precisely the area outside of Eastport, but the image had stuck. Planning a trip to the Bay of Fundy region in December, and hoping for a boat ride out to see the Beta TGU and the Energy Tide 2 vessel that deployed it, made me feel a bit seasick and as frozen as one of those fish sticks.

I shouldn't have worried. Weather conditions at the local reporting station at Pembroke, Maine, showed early December temperatures averaging in the 30s. Still mighty chilly for a boat ride, but compared to Laramie, a trip to the tropics.

I didn't know until I did some research on Maine that weather conditions along the coast are generally warmer than they are inland. In fact, there was a lot I didn't know about the ocean, having never spent much time around saltwater during my landlocked life. I grew up in Kansas and Missouri and have spent almost twenty years as a westerner. I'm used to seeing the ground everywhere I look, maybe interrupted by rivers and lakes, but not disappearing for good at the edge of the ocean.

They say that geography shapes perception, and I think they are right. For example, when I drove into Eastport over the Highway 190 causeway that connects two of the islands comprising the town, I saw beneath the bridge a flat muddy area, with water off in the distance. "Aha" I thought. "Drought!" I know too well that drought draws water down, like it has from our western reservoirs. But—aha—in other regions it is the lunar cycle that withdraws water and promptly puts it back.

Some people try to understand and harness the power of the

lunar cycle, the moon and the stars, by uttering incantations. "I can call spirits from the vasty deep" is an oft-quoted line from Shakespeare's Henry the Fourth, Part I, referring to one man's relationship to the ocean. Others, like ORPC based in Portland, Maine, use science to cast a spell on the moon to make it fall in love with the water, in a sense, and create electrical power as its love child. ORPC has brought its scientific methods and perseverance to the waters around Eastport. That's where the highest tides in the continental United States are found, with currents reaching about six knots, or seven miles per hour, at peak flows four times a day.

Eastport is located at the entrance to Cobscook Bay from Passamaquoddy Bay in the Bay of Fundy. That's a lot of bays to sort out, in addition to the body of water known as Western Passage that abuts them. A nautical chart shows clearly where one feature starts and the other stops. But standing on land, most observers are less picky about nomenclature.

There are also plenty of island names to get straight. Most of Eastport is on Moose Island, but bits and pieces of the town of about 1,600, during the high season, are scattered across Carlow Island, Spectacle Island, Goose Island, and Treat Island. Eastport claims to be the easternmost city in the United States, which it is. However, it only became that around the time of the millennium celebrations and the ensuing wrangling over what U.S. location would be the first to see the sunrise on the New Year. Rival community Lubec, which isn't far from Eastport by boat, actually juts out a bit further east than does Eastport. But Eastport beat Lubec to the punch, declaring itself a "city" and elevating itself above mere town status, leaving Lubec in the west.

Although I had planned for a December trip, by the time I worked through the various logistics of travel, it was mid-May before I actually went to Maine. I flew from Laramie to Denver to New York and eventually to Bangor. New York was a place I'd never been and so I found it impossible to picture from a geographical point of view just by looking at one-dimensional maps, even on three-dimensional Google Earth. My flight was through clear skies so I was able to see the Statue of Liberty, Lower Manhattan where the Twin Towers should have stood, and views of New York City that seem to appear in every movie ever made. I landed at LaGuardia Airport, which I never really left except to exit one terminal, hop a transit bus, and take a ten-minute ride to the next. But just putting my boots on the ground, locating the proper bus, seeing leaves flittering in the light breeze, hearing honking taxicab horns, and gobbling a slaw dog and mango drink from the Papaya King made me understand why eight million people want to live here. (Though they might not be the same eight million who actually do live there.) I think I would have enjoyed being a New Yorker in another era, sipping manhattans at a penthouse cocktail party hosted by famous-for-being-famous Kitty Carlisle. But until ORPC or someone else turns their research efforts to time travel, I think today's New York will have to get along without me.

After a commuter flight from LaGuardia I arrived in Bangor at nearly 11:00 p.m. I'd spent most of that two-hour flight observing a young man and woman, seatmates in a row across from me who hadn't previously met, pound down Bloody Marys and Budweisers, engage in flirtatious touching, and leave the plane promising to friend each other on Facebook. The woman declared to the whole plane that she'd "fallen sort of in love." What will she do with the boyfriend who works in the nuclear plant, as the whole plane couldn't help hearing of, or the fact that her new crush lives in Tennessee and she in northern Maine? We must leave that outcome to the maneuvers of the moon.

I fell asleep that night in Bangor pondering the star-crossed young couple and nearly overslept the next morning. Starting out at midday, I rented a car and took the long way from Bangor to Eastport. I aimed south, rather than the logical choice of east, and took Maine Highway 1, the coastal highway. How are there so many places I've never been, I wondered, as I drove through Bucksport, Ellsworth, Gouldsboro, Millbridge, Jonesboro, Machias, Whiting, and finally up to Pembroke, making the final turn off Maine Highway 1 at Perry. These towns aren't exactly household names, to be sure. But even this three-hour journey showed me more of Maine besides places I'd always heard of, like Portland, Bar Harbor, the Bush family residence at Kennebunkport, and Bangor, home to Stephen King.

When I reached Eastport in late afternoon I found my room at Motel East. I stood on my balcony and took in views across Western Passage to Campobello. The Roosevelt family had had a summer residence there for many years, and it is now an international park. International because Campobello is in New Brunswick, Canada. The lyrical word "Campobello" has been burned in my memory since I saw Ralph Bellamy starring as Franklin Delano Roosevelt in *Sunrise at Campobello* on late-night television. The movie tells the story of a young Roosevelt being afflicted by polio during a stay there. And there was the real Campobello before my eyes, just a few thousand yards across the water. Focusing closer in, I noticed seagulls sunning themselves on most available horizontal surfaces. I could see mud at the edge of the shore. Aha! Low tide!

Like any interested tourist, I laced up my sensible shoes, grabbed my camera, and commenced walking Eastport. I couldn't help noticing right away the folk-art-style, ten-foot-tall statue of a yellow-haired fisherman clutching a pollock to his chest. He wears a blue hat and shirt, tan pants, and tall yellow boots. He stands with his back to the "vasty deep," staring over rooftops of the downtown shops. I learned later he was built as an atmospheric detail for a murder-mystery/game-show/reality program. *Murder in Small Town X* was filmed in Eastport, which was dubbed "Sunrise" in the show. It ran for ten episodes in 2001. The winner of the show's game was a firefighter from the Bronx, New York, named Angel Juarbe Jr. A few weeks after the season finale, he and fellow responders entered the first of the Twin Towers just before it fell, on September 11. His body was not recovered. A plaque commemorating Juarbe's sacrifice to the nation is mounted on the base of the fisherman statue.

Walking on, I found an interpretive placard describing the downtown district, which is listed on the national register of historic places. Eastport's past includes its rise to prominence as a shipbuilding center and trading port. Major fires in 1839, 1864, and 1886 wiped out most structures, leaving the town with the comparatively modern Italianate architecture it has today. During the course of my walk I noticed at least one for-sale sign on every residential block, sometimes more. I wondered if the apparent buyer's market in town was a normal fluctuation, like the tides, or a sign of a deeper economic problem.

I got an answer from Sharon Cook at the *Quoddy Tides* newspaper office and wool shop. "Quoddy" is the local shorthand for Passamaquoddy, both the bay and the Indian tribe. The *Quoddy Tides* was established in 1968 and is the most easterly newspaper published in the United States. It comes out every other week and covers happenings in this part of Washington County, complete with the tide table and the shipping news.

Cook, with pale red hair and a smooth round face, has lived in Eastport all of her sixty years. She's been to many of the neighboring islands but not all. "I don't like boats," she explained. She's seen a slow steady decline in Eastport's population and an increased reliance on tourism and boutique shops for the town's economic base. Most of the people moving in are retirees or seasonal residents. "There aren't many jobs for people these days since the sardine canning factories moved inland," she said. "Are you looking for a house?" she asked me. "My mother's is for sale."

I wasn't looking for a house but I *was* looking for a seafood dinner. Few places are open in Eastport before the summer season starts on Memorial Day. But luckily I found the Happy Crab, just a few doors down from my motel.

It seemed that everyone in Eastport had just gotten off work and stopped into the Happy Crab for beverages or an early dinner. I took the only empty seat where I soon found myself in conversation with a lady of indeterminate age sporting the black hair and cropped bangs of a silent-film star. Nan Milani introduced herself and pumped me for a few personal credentials. Then she introduced me to the "gang." I felt like I was at one of Kitty Carlisle's New York parties. "This is Julianne Couch. She teaches at the University of Wyoming and she's here to write about tidal power." Then she introduced me to folks including Dean Pike, (owner of the Moose Island Marina); Stan from Oklahoma (pharmacist for the Passamaquoddy Reservation, Pleasant Point, Eastport); Bob Del Pappa, aka Bar Harbor Bob (owner of The Chowder House Restaurant and former deep-sea diver). And of course, herself. Nan Milani ("Past President, State of Maine Federation of Republican Women," as she explained it to me). We sat chatting about books, the New York opera scene, and New England history as we ate dinner. Before I knew it I had three business cards, two sets of addresses on cocktail napkins, and several invitations to return for the annual Fourth of July event, featuring the largest parade in Maine. "I'm a connector," she told me, as if I needed any more proof.

With a belly full of crab roll and Budweiser wheat beer, which kept showing up in front of me as if by magic, I was almost relieved to see happy hour come to an end. After all, I was here to work, not socialize. I said my farewells and headed to my motel room to watch the tide come in, and to study up on ORPC's efforts to harness the moon.

I was looking forward to the next morning, when I'd meet ORPC marketing manager Susy Kist, with whom I'd been communicating by e-mail, and director of operations Bob Lewis. Kist is a Chicago-area native who'd found her way to Maine as many lovers of the outdoors do: through its hiking trails. She'd worked many years in the nonprofit sector as a special event coordinator, fundraiser, and executive director for environmental and cultural organizations, before coming to ORPC in 2009. Her colleague Bob Lewis is an Eastport native who has been with ORPC since 2007. He has master's degrees in chemistry and business administration. He manages ORPC's tidal energy projects, including planning, budgeting, and supervising site operations.

ORPC's president and CEO was a cofounder of the company back in 2004. Christopher R. Sauer earned a civil engineering degree from the University of Illinois. He has more than thirty-five years of experience in industrial engineering, energy project management, and startup company formation. When he describes the work his company does, he says: "We're in the ingenuity business."

According to Sauer,

> ORPC has been hard at work developing breakthrough technology and eco-conscious projects that use ocean and river energy to produce clean, predictable, affordable energy to power homes and businesses while protecting the environment. We believe that ocean and river power systems not only can, but must, be designed to preserve the world's marine ecosystems. They should provide renewable supplies of clean energy that are both reliable and affordable. They should be compatible with other marine industries, and should create jobs and other economic opportunities in their surrounding communities.

The company has several proprietary tidal power technologies in development that they hope reach these goals. One is the TidGen Power System, designed to generate electricity at water depths of fifty to one hundred feet. The company says this system is used at shallow tidal and deep river sites. Groups of TGUs will connect directly to an on-shore substation through a single underwater transmission line. In this system, each TGU generates up to 250 kilowatts in a six-knot water current (with 1,000 kilowatts comprising a megawatt).

ORPC's OcGen Power System, the largest and most powerful of their modular systems, is designed for use in water depths of more than eighty feet. In the OcGen Power System, up to four TGUs may be stacked together to create larger power-generating modules that are moored to the sea floor with a low-impact mooring system designed specifically for this purpose. Anywhere from a few to several dozen modules will be located at the same site, and will be connected in groups to an on-shore substation through a single underwater cable. A module composed of four TGUs will have a peak generating capacity of 1,000 kilowatts in a six-knot water current, according to the company.

The last of ORPC's power systems is called RivGen. It will generate electricity at small river sites, particularly in remote communities with no large, centralized power grid. Currently, many of these communities rely on local power-distribution grids connected to diesel generators, which leave a significant carbon footprint and are growing increasingly expensive to operate. The RivGen Power System is designed to connect directly into these existing diesel-electric grids and to provide automatic fuel switching so that whenever the RivGen Power System is generating power, the diesel generator turns down or off. Depending on community needs and site size, the RivGen Power System can include up to several dozen TGUs, with each TGU generating up to 30 kilowatts in a ten-foot-per-second river current. I was in Maine to see ORPC's beta turbine generator unit (Beta TGU), the precursor to ORPC's TidGen, OcGen, and RivGen technologies.

The sequence of steps leading to the research, development, and deployment of these systems is methodical, by necessity. Susy Kist explained where they were in the sequence. "We've done the prototype TGU, now we're doing the beta version. Our next step will be getting a TidGen Power System in the water and connected to the grid in 2011. An OcGen Power System will follow, and then RivGen. These are not fully developed yet. We're still working on the building blocks of the technology."

As part of the permitting process, ORPC developed and tested a prototype TGU in Western Passage. Once they were able to show that the prototype could create electricity, they were ready to test a precommercial version of the Beta TGU, to prove that their proprietary technology would produce grid-compatible power. After my visit, in the first half of 2012, following approval by the Federal Energy Regulatory Commission (FERC), ORPC began the Maine Tidal Energy Project by installing a commercial TidGen Power System in Cobscook Bay. After running and monitoring this initial system for a year, they will install additional power systems over the ensuing three years to increase the project's capacity to 3 megawatts—enough electricity to power 1,200 Maine homes and businesses with clean tidal energy.

As they advance their technology, ORPC has made good on Sauer's goal of creating jobs and economic opportunities for the regions where the company is working. For example, the Energy Tide 2 barge was built by Morrison Manufacturing in Perry, Maine. The deployment system for the TGU was built by Alexander's Welding and Machine in Greenfield, Maine. The deployment system was mounted to the Energy Tide 2 in Eastport.

Fabrication of the 100-percent-composite turbine foils was done in Bath, Maine, by U.S. Windblade. Turbine foils are like rotors on a windmill, only instead of spinning in the air, they rotate under water. Composite parts of the steel and composite TGU frame were made by Harbor Technologies of Brunswick, Maine. Steel parts of the TGU frame and overall assembly of the frame was by Stillwater Metalworks in Bangor. The generator itself was built in Marlboro, Massachusetts. All of these parts were delivered at various times to Eastport and assembled at a local technical college called The Boat School. Then the Energy

Tide 2 research vessel was brought to The Boat School ramp, and it and the TGUs were attached to the deployment system. The whole affair was pushed by an Eastport Port Authority tugboat to the project mooring site in Cobscook Bay. Perhaps in acknowledgement of these economic developments, the city of Eastport and Washington County received a $1.4-million award to establish the Maine Renewable Energy Manufacturing facility.

I awakened briefly at about 4:00 a.m. the morning of my meeting with Kist and Lewis, aware of being one of the first people in the country to see the sunrise, over Campobello. A few hours later I made the short walk to ORPC's office, which is one-half of what used to be a grocery store and is now among the best office spaces in town. Kist had made the six-hour drive from ORPC's headquarters in Portland, Maine, the night before. She loaded me up into her car then we headed out to see the sights, making our initial stop a mercy visit to the IGA grocery where I could get my morning coffee.

From there we drove to The Boat School, the nation's oldest boat-building school, "located where the wilderness meets the sea." Scupper, a calico cat named after a drain on a boat deck, greeted visitors to the office. Administrative assistant Caryn Vinson gave us a quick tour. Boat building is a four-hundred-year-old Maine tradition, she told us. The Boat School offers training in wood construction, marine systems, composite and fiberglass construction, and marine finishes to keep that tradition alive. We didn't have to wear safety equipment when we entered the shop space because the school had just graduated a class of students and was in its slack time between sessions. We moved from a large shop where students build wood vessels to a classroom where students take their English course. Most of their writing assignments are on the subject of boats or water. Vinson told us that the mostly male, traditional college-aged graduates of their two-year program usually get jobs right away. "They get jobs that aren't advertised, because of the school's reputation for turning out quality graduates. One boatyard didn't have a position open, but they told us, 'Just send us your best graduate and we'll find a place for him.'"

Before starting at The Boat School, Vinson had been the librarian at the local elementary school for thirteen years until her position was eliminated. "It broke my heart when the library closed due to budget cuts, but life goes on, and I absolutely love the job I have now," she told me. "I cannot imagine working in a more positive or a more fascinating place. I learn something new every day."

Our next stop was The Boat School's ramp where the first prototype TGU was launched. It is now out of the water and stored nearby under a tarp. The prototype TGU is about twenty-five-feet long, about half the size of the next model they developed, the Beta TGU. Kist put gentle pressure with her hand against the black foils of the prototype to show me their movement against the force of water.

Some people are concerned that anything in the water with motion reminiscent of a push lawnmower might present a hazard to fish. The University of Maine's School of Marine Sciences has been working with ORPC to understand and mitigate any impact of this technology on wildlife. "There's been obvious concern that when you put a turbine in the water, there could be potential impact," said Gayle Zydlewski, a fish biologist with the University of Maine's School of Marine Sciences. "If we do it right, it can be done in a safe and sustainable way."

According to Christopher Sauer, "To date, our turbine generator unit testing has caused no negative effects on either fish or sea mammals, which tend to simply avoid the unit the way they would a rock or other natural barrier, by swimming around it. Since the turbine foils rotate slowly and do not funnel or suck water into them, they pose minimal risk to the fish that do swim through the units."

At a tidal power forum sponsored by the Tidal Energy Demonstration and Evaluation Center at Maine Maritime Academy,

Dana Murch, a hydropower specialist with the Maine Department of Environmental Protection, noted that there is little information about how these developing technologies affect the environment and existing uses such as recreational boating and commercial fishing operations. "There's no place to go to tell us what these effects are," he said. "There's no discussions out there anywhere in the world."

Anne Miles, the director of the hydropower licensing division at the Federal Energy Regulatory Commission, was the keynote speaker at the tidal power forum. She indicated that ORPC's research is vital to understanding long-term effects of tidal power systems on marine life. "We're learning by getting projects in the water," she said.

Another important goal for ORPC that coincided with my visit was to gather data during one full lunar cycle showing that the energy created by the Beta TGU is grid-compatible. This process of being licensed by FERC to put power onto the electrical grid is a long one in which energy developers have to prove they know what they are doing and can do what they say they can. To give me a better understanding of the Beta TGU's potential, Kist and Bob Lewis had arranged for me to board a boat and head out for the "Icy North Atlantic."

Not literally out to the open ocean where fish sticks are netted, of course. We never left Cobscook Bay or Western Passage. Our orange life vests secured, veteran captain Skip Harris gave Kist and me an hour-long ride on his cousin Butch's converted lobster boat, the *Lady H*. "It's been Coast Guard certified to take passengers," Harris assured us. The waters here are not only cold (about 52 degrees at the time of our trip), they are deep. It was late morning and clouds were beginning to roll in. It felt good to stand near the boat heater and warm our hands, watching the scenery of green islands and blue water roll and bob past the windshield. We motored around to see the waters near Dog Island in Western Passage where the prototype TGU was once deployed. Then we were off across Cobscook Bay where the Beta TGU is now deployed. Harris and Kist pointed out landmarks along the way in Maine and in New Brunswick.

Harris, and his cousin Butch, are from a Maine family that goes back enough generations that several landmarks are named for them, such as Harris Cove, which we passed. He pointed out "the Old Sow," the largest tidal whirlpool in the Western Hemisphere, located between the southwestern shore of Deer Island in New Brunswick, and Moose Island. Harris had a well-worn copy of *Peterson's Field Guide to Eastern Birds* on the dash of the boat to identify various birds and waterfowl. He didn't need to use it to identify the several bald eagles we observed soaring over the water in search of fish.

Harris was in the Coast Guard for a number of years before he became the captain of a swordfish boat in Martha's Vineyard. He'd only recently returned to his native Eastport to live. I noticed that he wore a hearing aid and over the roar of the *Lady H*'s diesel engine I asked him if too much time on loud boats was to blame. "All those years working in the engine rooms on ninety- and one-hundred-foot boats without hearing protection—I'm paying for it now," he shouted back.

After about fifteen minutes we closed in on the Energy Tide 2 vessel, moored to a position where the Beta TGU would be held in a fairly constant location. Kist explained that the Energy Tide 2, and therefore the Beta TGU, swing with the tide during this testing phase. The waters here are 160 feet deep, Harris told us. The location for the placement of the Energy Tide 2 was one that fishermen and others in the community had approved because it would not interfere with their commercial or recreational activities.

At the time of our visit, the Beta TGU was out of the water, in its "up" position. Earlier testing had revealed that more "tweaking" of the generator would be required. So it was removed and sent to Massachusetts, where it was constructed, for additional improvements. That was unfortunate for ORPC because it meant no data were being collected. But it was fortunate for me because I got to see the hydraulic arms that raise and lower the Beta TGU into the water, the blue-and-white turbine foils, and the gap in the middle where the generator would go. The TGU technology uses an underwater, permanent magnet generator.

When the Beta TGU is commercially operational and putting power onto the grid, it will no longer be deployed by the Energy Tide 2 barge, Kist explained. Instead, it will be secured to the ocean floor by a bottom support frame. It is likely that several turbines will be connected to make the most of the movement of the tides.

The Energy Tide 2 has more to do than simply dangle the Beta TGU into the vasty deep. It is a well-equipped scientific vessel on which researchers are stationed to collect data about how much power is being created from the tidal energy. Tides come in and they go back out. But they also stop in order to reverse directions, in a phase known as slack tide. Bob Lewis had explained that although power can't be produced during these times, at least tides are predictable. "You can't always know when the wind will blow or the sun will shine," he said in a comparison to wind and solar power. "But we can predict the tides."

Skip Harris took the *Lady H* on a couple passes around the Energy Tide 2, and the *Anna Frank*, a smaller lobster boat, which had brought a few of the ORPC's staff on board. We waved to those fellows, Dave Turner and Ryan Beaumont, and headed back toward Eastport. First, Harris offered to take us closer to the bridge at the town of Lubec that crosses into Campobello. Kist had been explaining to me that the trip to Lubec would take just a few minutes by boat but more like an hour by car. I kept repeating dumbly what she said, unable to understand why a bridge I could hit with a well-aimed fourteen-pounder from our location in the water could be an hour away by land. From all his years guiding swordfishermen, Harris must have known my type, so southward we went toward Lubec to let me see the scenic town's location on the shoreline for myself.

Along the way they told me about the annual Pirate Festival, in which citizens from Eastport dress as pirates, board all available boats including Skip's cousin Butch's eighty-four-foot schooner the *Sylvania H. Beal*, and launch a tomato-fueled attack on Lubec. They explained the history of piracy in the area that the event marks. After the American Revolution, the bay area had an active smuggling trade. At one point, flour was smuggled from American territory into New Brunswick. Then during the War of 1812, British manufactured goods saw a brisk illegal business. Later, gypsum from Nova Scotia was the hot item. Harris indicated that other contraband such as liquor might have passed this way at one time, too.

I was already planning my Olivia de Havilland costume for the next Pirate Festival—even though her pirate films, like *Captain Blood* with Errol Flynn, were set in the Caribbean, but what's a little geographic detail when there's a chance to attack a village by sea?—when Harris brought us back to the Eastport pier. He slid the *Lady H* into a spot that seemed barely large enough for a crab pot. Kist and I said "so long" to him and headed to Quoddy Bay Lobster, right on the shore. Over crab rolls and bottles of fresh water, we talked about what we'd learned that day.

Bob Lewis had spoken to us that morning of what the ORPC was doing in aviation terms. "We are like the Wright brothers when they learned to fly at Kitty Hawk." No one has done this sort of work before and it takes a great deal of effort and patience, he said. Most of us can't imagine a world without airplanes today, thanks to the Wright brothers. "Perhaps one day we won't be able to imagine a world in which polluting or disaster-prone forms of power are the norm," he added.

ORPC's power systems produce no emissions and require no fossil fuels to operate, deriving their power solely from the renewable resource of the earth's oceans and rivers. That makes them highly attractive, by most anyone's definition. But hydrokinetic energy is only part of the answer. The Electric Power Research Institute estimates that at the sites it has studied—in areas with powerful tides like easternmost Maine, the Pacific Northwest, and Alaska—a total of about 13,000 megawatts is potentially available. As of now, Lewis is reluctant to pinpoint the capacity of what ORPC could produce around Eastport, just like he's reluctant to speculate about efficiency, which he says is currently being discussed in "apples and oranges terms."

Apples and oranges is a good way to consider capacity and efficiency, and might be a good way to think about power consumption in the state of Maine, with apples being the sort of power sources they do have and oranges the sort of power they could have. With its population of 1.3 million distributed among a few midsized cities and widely flung rural communities, Maine has unique energy needs. The bulk of the population lives in southern Maine, in and around its largest city, Portland. Most of Maine's communities use fuel oil for heating. Shipments of crude oil come through its ports and are sent to refineries in Quebec and Ontario, since Maine has no refineries of its own. While long winters in Maine mean lots of furnaces running, cool summers mean Mainers do not use much electricity for air conditioning. Maine has no fossil-fuel reserves but has substantial renewable energy potential. "The state's numerous rivers, forests, and windy areas provide the potential for hydroelectric, wood-fired, and wind-powered generation," according to the Energy Information Administration.

That potential led Maine's Public Utilities Commission to adopt in 1999 a renewable portfolio standard requiring that at least 30 percent of retail electricity sales come from renewable sources. In June 2006, Maine adopted another renewable portfolio goal to increase renewable energy capacity by 10 percent before 2017. While some states establish these goals as strong suggestions, Maine made its renewable capacity goal a mandatory target in 2007. At ORPC's launching ceremony for the Beta TGU and the Energy Tide 2, former Maine governor Angus King said the idea of tidal power isn't new, but it is sensible. "It's as if we had a gigantic pool of oil under us. All we have to do is tap into it."

At the picnic table, over the remains of our lunch, Kist and I put our jackets back on as the sun disappeared for the day behind an approaching weather front. We talked more about other types of renewable energy, as well as the country's current consumption of oil and coal. Kist's father is a retired engineer who worked in the nuclear power industry. She said she felt somewhat conflicted about that form of power because of its disadvantages, such as safe long-term storage of radioactive waste. The bottom line for her, she said, was that whatever form of power goes out onto the grid or whatever type of fuel goes into our engines, we should use less of it.

Kist is proud of the work the ORPC is doing and is excited for the day when the power will go onto the grid. That would make the investors happy and help people in Eastport breathe a relieved sigh, knowing that their town could get back on solid footing with a steady economic partner like ORPC. One business benefiting indirectly from ORPC's activity is Moose Island Marine, owned by Dean Pike, to whom Nan Milani had "connected" me at the Happy Crab. In addition to operating the marine store, Pike was trained as a marine biologist and teaches at The Boat School. When I met him he was wearing a blue, long-sleeved T-shirt that said: "Get Your Ship Together at the Port of Eastport, Maine." A blue baseball cap with his business logo covered his head. He told me he used to have "wicked long hair" but when he started to go bald he began to look like Bozo, so he cut it all off. Pike said he has been "messing around" with boats for about forty years now, and loves everything about Eastport except January, February, and March. He vouches for the economic impact the work of ORPC has made in the community. "I lost less money over the winter than I ordinarily lose. The ORPC being here made it so that I didn't have to close down for the winter season."

This is not the first time an idea for using Eastport's tidal resource has gotten the community excited and optimistic, but Kist hopes the ORPC's project has a better outcome than the last project did. Though most area residents don't remember the failed Quoddy Dam project during Roosevelt's administration, they can learn all about it by visiting the craft shop where the model of the Quoddy Dam project is housed. Bob Lewis took Kist and me there after our crab-roll lunch. The model was built in the 1930s to show where the tides would have been captured in a barrage system. A "barrage" is basically a barrier, similar to a dam, with turbines mounted in it. It holds back water in high tide then releases it all at once. The high velocity of the rushing water turns the turbines and stimulates the generator. However, the barrages affect the natural movement of saltwater within an estuary, with a negative impact on ecosystems. Environmental objections to this type of system are voiced loudly today, though in the 1930s the drawbacks were not fully understood.

In the 1930s, a barrage system seemed a fine way to capitalize on the deep waters and strong tides around Eastport, Lewis told us. It all started when a developer named Dexter Cooper got the idea to use the tides around Eastport to create cheap electricity. He presented his idea to various private power concerns, but they all rejected it. So did the Public Works Administration, part of Roosevelt's administration. But then Cooper played the childhood friendship card, since his wife knew the Roosevelt family from childhood summers in Campobello. Cooper went directly to Roosevelt with his idea. By this time, the federal government was spending significant amounts of money on New Deal policies, designed to put people to work and end the Great Depression. Roosevelt liked Cooper's idea and allocated a budget of $10 million to make it happen.

According to a *March of Time* newsreel from the era, on

July 4, 1935, Eastport celebrated the news of the deal with a Fourth of July parade unlike any other in size and festivity. Thousands of Maine's unemployed workers began pouring into Eastport to get jobs on the project. As the baritone narrator of the newsreel announced, the relief lines were gone. "Day after day more and more men pour in, signing up for the task of harnessing the moon."

The Army Corps of Engineers became involved, basically rejecting Cooper's original plans and instead planning their own barrage system. The intended result was "a scheme to bottle up Cobscook Bay completely," so that the little islands dotting the bay would be connected together to form a reservoir of thirty-seven square miles. High tide would flow inland and be trapped by the barrage gates. When the tide went out and dropped several feet below the level of the impounded water, the gates to the powerhouse would open, water would flow past and through the turbines, generating electricity before heading back out to sea. A model was built to demonstrate how the system would work. That's the model Bob Lewis and some other Eastporters dusted off, repaired and repainted, and placed on display in the craft shop.

"It was the private power companies that killed the idea," he told Kist and me. He explained that Maine did not have a tradition of public utilities in the state, and the private utility owners were afraid of turning over their lucrative industry to public interests. The newsreel reinforces this idea, describing the initial excitement among Maine Republicans for the ideas of a Democrat president that could make Eastport a prime seaport of the North Atlantic and the "nerve center" of Maine's power supply, but which ended as a "monument to disappointment."

Indeed, work began on the project, with rock being moved to build causeways connecting the islands, such as Highway 190, which I'd taken into town from Perry. But soon Roosevelt's New Deal plans, including the Quoddy Dam project, were under attack in Congress. Cost overruns and loud objections by the power companies that there were too few customers in northeastern Maine to make the project financially viable were the chief issues. Another was that to keep the electricity flowing during the two periods of slack tide daily, an auxiliary power system would have to be built. So a year after it began, Congress voted to cancel the Quoddy Dam project. As the newsreel narrator put it, Mainers rushed back out of Eastport like the tides, "jobless again, thumbing their way back to relief." Instead of working on the dam project, they were shown unloading pots from lobster boats, "returning to dependable ways of getting a living from the sea. Time marches on!"

Time is marching on for ORPC, the people of the Eastport area, and indeed for everyone seeking to find alternatives to unsustainable forms of power. The hard work of siting, permitting, developing, and championing new ways to create electrical power might strain the patience of some people. It helps to have the support of the local community and people in positions of authority and influence around the state.

At the same as the ORPC launch ceremony where Angus King spoke, Maine state senator Kevin L. Ray discussed ORPC's potential impact on the area. He said this project could result in "many new jobs, something this area desperately needs." He praised ORPC's effective, community-based approach for siting and designing projects and their willingness to work with others who use the water "to minimize and avoid conflicts."

His comment about avoiding conflicts was an understatement. The ocean has long been a battlefield. Since humans first learned to build boats to defend their turf in the surf, wars have been fought on water. The pirates who smuggled goods between the small islands of the Bay of Fundy knew it. The regular military who fought the War of 1812 knew it. In fact, the town of Perry, Maine, was named for the commander of the American fleet who defeated the British fleet in the Battle of Lake Erie. On September 10, 1813, Oliver Hazard Perry wrote some to-be-famous words in a letter to Maj. Gen. William Henry Harrison, after defeating the British fleet in that battle. "Dear Gen'l: We have met the enemy, and they are ours, two ships, two brigs, one schooner and one sloop."

I'd often heard the first part of that quotation but didn't know its origin. The original words had already morphed in my mind to the expression: "We have met the enemy and he is us." That version was made famous in the cartoon strip *Pogo*, drawn by Walt Kelly. Main character Pogo, his cigar-smoking alligator pal, Albert, and the rest of their friends lived in hollow logs and such in Georgia's Okefenokee Swamp. Kelly was said to have reclaimed that quotation from Perry as an attack on McCarthyism in the 1950s. He used it again in an environmental poster he created in 1970 to mark the first celebration of Earth Day, the one Ira Einhorn helped found.

While we still have skirmishes at sea, the most recent battle over the ocean is an environmental one. In some locations, towering wind farms are being erected offshore, capturing the steady ocean breezes. Opponents of this technology do not like their viewscapes cluttered by windmills. They make arguments just like those heard in Wyoming about siting wind farms in people's backyards, so to speak. Tidal projects that are variations on what the ORPC is doing are being considered by other companies for the waters off Florida, Alaska, and internationally. Some use barrage systems, others deploy turbines shaped like windmills swirling in the water. The environmental impacts of these new technologies are as yet unknown.

By contrast, the environmental impacts of other technologies are known and have been experienced in disastrous doses. When the Exxon Valdez leaked 10.8 million gallons of oil into Alaska's Prince William Sound in 1989, we watched biologists and volunteers daub oil off ducks with Dawn soap and Q-tips. In the Gulf of Mexico, offshore oil drilling led to a devastating industrial disaster in 2010, when a rig exploded and a volcano-like eruption of crude oil floated to the surface and globbed its way to shore. There might not be enough Dawn and Q-tips in the world to scrub up from that eruption of the Deepwater Horizon oil rig, operated by Transocean as a contractor for BP. Creatures of both sea and land, like in Walt Kelly's Okefenokee Swamp, must feel that they, too, have met the enemy, and he is us.

To get a final view out into Cobscook Bay my last morning in Eastport, I took a short hike at Shackford Head trail. Interpretive signs in the trail parking area explained that here at Cony Beach, five Civil War ships that served the Union Navy were burned, between 1901 and 1920. This was so the several hundred tons of precious resources like copper, brass, and iron in their hulls could be salvaged. The ships were brought to this area so that during high tide they could float to shore. Work commenced when the tide receded, leaving the ships in place. Before the ships were torched, items such as furniture, wooden doors, and staircases were sold to locals. Some of these items are still in Eastport homes, today.

I left the interpretive area and hiked a short way through deciduous and pine trees, up past lush ferns, and what I hoped wasn't poison ivy. At the top of the lookout was a conveniently placed bench. Resting there I could see about a dozen aquaculture pens where Atlantic salmon are raised. In aquaculture, fish are raised and artificially fed in pens that float in the water. According to the National Oceanic and Atmospheric Administration, about half the fish eaten by humans in the world today are raised commercially this way, rather than taken as wild creatures from the sea.

Humans are harnessing the power of so much already, from fish to air to water to atoms, to the liquids and gases in the earth, to the heat of the sun, to the combustibility of coal and woody fuels. So why not harness the moon? As they say in my native Kansas, *Ad astra per aspera.* To the stars through difficulties. With intelligence, ingenuity, and supportive public policy, ORPC and others can someday produce sustainable, renewable power on a commercial scale at reasonable cost without environmental damage. That might sound like magic—and if we define that term as science plus serendipity, that's what it is. We can do our part by consuming less and developing and distributing more of our own power locally. We can take the responsibility to better analyze the information presented to us by boosters of various causes, knowing the difference between numbers meant to impress, stories meant to persuade, and facts that prompt action. Finally, we can remember that North Americans are not alone on the planet, that our actions invite reactions from others, and that cooperation is worth a try to keep our world inhabitable for all its residents.

Afterword

In the course of my travels along the power line, I happened upon something I hadn't realized I was looking for. I discovered a town in eastern Iowa on the banks of the Mississippi that 2,300 people already had claimed as home. There I found a 130-year-old Victorian house, in good condition but in need of restoration sooner rather than later. It was just the right challenge for my husband and me. We decided to give it a go. Now we face the hurdles of bringing this nineteenth-century home into the twenty-first century. That will mean remodeling with sustainable technologies in mind—from renewable materials like cork for the flooring, Energy Star appliances, low-flow toilets, energy-efficient windows, and better insulation all around. It'll also mean no more smugness about not having central air conditioning inflating our carbon footprint. If air conditioning was invented for any location, it might have been for hot, humid Midwest summers. But I pledge to keep the thermostat set to 78 in summer and to spend most of my days outside on the shaded balcony, watching barges carrying cargo such as coal and corn through Lock and Dam 12. I'm hoping to hitch a ride and find out how river transportation works.

A Note on Sources

To research this book, in addition to travel and interviews, I relied on many books, reports, articles, and online information. These are noted within the text. Among the most valuable were:

Beder, Sharon. *Power Play: The Fight to Control the World's Electricity.* New York: The New Press, 2003.

Bodanis, David. *Electric Universe: How Electricity Switched on the Modern World.* New York: Broadway Books, 2006.

Clemens, Kevin. *The Crooked Mile: Through Peak Oil, Biofuels, Hybrid Cars, and Global Climate Change to Reach a Brighter Future.* Lake Elmo MN: Demontreville Press, 2009.

Commercial Wind Energy Development in Wyoming: A Guide for Landowners. 2nd ed. Laramie: University of Wyoming, 2011.

de la Garza, Amanda, ed. *Biomass: Energy from Plants and Animals.* Fueling the Future. Farmington Hills MI: Greenhaven Press, 2007.

Deffeyes, Kenneth. *Beyond Oil: The View from Hubbert's Peak.* New York: Hill and Wang, 2005.

Emerich, Monica. *The Gospel of Sustainability: Media, Market and LOHAS.* Champaign: University of Illinois Press, 2011.

Energy Information Administration. *Annual Energy Outlook 2011.* http://www.eia.gov.

Ezzell, Patricia Bernard. *TVA Photography, 1963–2008: Challenges and Changes in the Tennessee Valley.* Jackson: University Press of Mississippi, 2008.

Hayes, Brian. *Infrastructure: A Field Guide to the Industrial Landscape*. New York: W. W. Norton, 2005.

Hills, Richard L. *Power from Wind: A History of Windmill Technology*. West Nyack NY: Cambridge University Press, 1994.

Kolbert, Elizabeth. *Field Notes from a Catastrophe: Man, Nature, and Climate Change*. New York: Bloomsbury, 2006.

Krigger, John, and Chris Dorsi. *The Homeowner's Handbook to Energy Efficiency: A Guide to Big and Small Improvements*. Helena MT: Saturn Resource Management, 2008.

Lauglin, Robert. *Powering the Future*. New York: Basic Books, 2011.

Lovelock, James. *The Revenge of Gaia: Earth's Climate Crisis and the Fate of Humanity*. New York: Basic Books, 2007.

MacKay, David J. C. *Sustainable Energy—Without the Hot Air*. Cambridge UK: UIT Cambridge, 2008.

Marris, Emma. *Rambunctious Garden: Saving Nature in a Post-Wild World*. New York: Bloomsbury, 2011.

Makhijani, Arjun. *Carbon-Free and Nuclear-Free: A Roadmap for U.S. Energy Policy*. A Joint Project of the Nuclear Policy Research Institute and the Institute for Energy and Environmental Research. Oakland CA: IEER Press and RDR Books, 2007.

McPhee, John. *Uncommon Carriers*. New York: Farrar, Straus and Giroux, 2006.

Molvar, E. M. *Wind Power in Wyoming: Doing It Smart from the Start*. Laramie WY: Biodiversity Conservation Alliance. Available online at http://www.voiceforthewild.org/blm/pubs/WindPowerReport.pdf.

Natural Resources Defense Council. http://www.nrdc.org.

Nuclear Regulatory Commission. http://www.nrc.gov.

Rosenberg, Norman J. *A Biomass Future for the North American Great Plains*. Berlin: Springer, 2007.

Sobey, Ed. *A Field Guide to Roadside Technology*. Chicago: Chicago Review Press, 2006.

Union of Concerned Scientists. http://www.ucsusa.org.

U.S. Energy Information Administration. http://www.eia.doe.gov.

U.S. Nuclear Regulatory Commission. *2010–2011 Information Digest*. NUREG 1350. Vol. 22. Available online at http://www.nrc.gov.

In the Our Sustainable Future series

Ogallala: Water for a Dry Land
John Opie

Building Soils for Better Crops: Organic Matter Management
Fred Magdoff

Agricultural Research Alternatives
William Lockeretz and Molly D. Anderson

Crop Improvement for Sustainable Agriculture
Edited by M. Brett Callaway and Charles A. Francis

Future Harvest: Pesticide-Free Farming
Jim Bender

A Conspiracy of Optimism: Management of the National Forests since World War Two
Paul W. Hirt

Green Plans: Greenprint for Sustainability
Huey D. Johnson

Making Nature, Shaping Culture: Plant Biodiversity in Global Context
Lawrence Busch, William B. Lacy, Jeffrey Burkhardt, Douglas Hemken, Jubel Moraga-Rojel, Timothy Koponen, and José de Souza Silva

Economic Thresholds for Integrated Pest Management
Edited by Leon G. Higley and Larry P. Pedigo

Ecology and Economics of the Great Plains
Daniel S. Licht

Uphill against Water: The Great Dakota Water War
Peter Carrels

Changing the Way America Farms: Knowledge and Community in the Sustainable Agriculture Movement
Neva Hassanein

Ogallala: Water for a Dry Land, second edition
John Opie

Willard Cochrane and the American Family Farm
Richard A. Levins

Down and Out on the Family Farm: Rural Rehabilitation in the Great Plains, 1929–1945
Michael Johnston Grant

Raising a Stink: The Struggle over Factory Hog Farms in Nebraska
Carolyn Johnsen

The Curse of American Agricultural Abundance: A Sustainable Solution
Willard W. Cochrane

Good Growing: Why Organic Farming Works
Leslie A. Duram

Roots of Change: Nebraska's New Agriculture
Mary Ridder

Remaking American Communities: A Reference Guide to Urban Sprawl
Edited by David C. Soule
Foreword by Neal Peirce

Remaking the North American Food System: Strategies for Sustainability
Edited by C. Clare Hinrichs and Thomas A. Lyson

Crisis and Opportunity: Sustainability in American Agriculture
John E. Ikerd

Green Plans: Blueprint for a Sustainable Earth, revised and updated
Huey D. Johnson

Green Illusions: The Dirty Secrets of Clean Energy and the Future of Environmentalism
Ozzie Zehner

Traveling the Power Line: From the Mojave Desert to the Bay of Fundy
Julianne Couch

To order or obtain more information on these or other University of Nebraska Press titles, visit www.nebraskapress.unl.edu.